Communications in Computer and Information Science 2169

Editorial Board Members

Joaquim Filipe ⓘ, *Polytechnic Institute of Setúbal, Setúbal, Portugal*
Ashish Ghosh ⓘ, *Indian Statistical Institute, Kolkata, India*
Lizhu Zhou, *Tsinghua University, Beijing, China*

Rationale

The CCIS series is devoted to the publication of proceedings of computer science conferences. Its aim is to efficiently disseminate original research results in informatics in printed and electronic form. While the focus is on publication of peer-reviewed full papers presenting mature work, inclusion of reviewed short papers reporting on work in progress is welcome, too. Besides globally relevant meetings with internationally representative program committees guaranteeing a strict peer-reviewing and paper selection process, conferences run by societies or of high regional or national relevance are also considered for publication.

Topics

The topical scope of CCIS spans the entire spectrum of informatics ranging from foundational topics in the theory of computing to information and communications science and technology and a broad variety of interdisciplinary application fields.

Information for Volume Editors and Authors

Publication in CCIS is free of charge. No royalties are paid, however, we offer registered conference participants temporary free access to the online version of the conference proceedings on SpringerLink (http://link.springer.com) by means of an http referrer from the conference website and/or a number of complimentary printed copies, as specified in the official acceptance email of the event.

CCIS proceedings can be published in time for distribution at conferences or as post-proceedings, and delivered in the form of printed books and/or electronically as USBs and/or e-content licenses for accessing proceedings at SpringerLink. Furthermore, CCIS proceedings are included in the CCIS electronic book series hosted in the SpringerLink digital library at http://link.springer.com/bookseries/7899. Conferences publishing in CCIS are allowed to use Online Conference Service (OCS) for managing the whole proceedings lifecycle (from submission and reviewing to preparing for publication) free of charge.

Publication process

The language of publication is exclusively English. Authors publishing in CCIS have to sign the Springer CCIS copyright transfer form, however, they are free to use their material published in CCIS for substantially changed, more elaborate subsequent publications elsewhere. For the preparation of the camera-ready papers/files, authors have to strictly adhere to the Springer CCIS Authors' Instructions and are strongly encouraged to use the CCIS LaTeX style files or templates.

Abstracting/Indexing

CCIS is abstracted/indexed in DBLP, Google Scholar, EI-Compendex, Mathematical Reviews, SCImago, Scopus. CCIS volumes are also submitted for the inclusion in ISI Proceedings.

How to start

To start the evaluation of your proposal for inclusion in the CCIS series, please send an e-mail to ccis@springer.com.

Bernhard Moser · Lukas Fischer · Atif Mashkoor ·
Johannes Sametinger · Anna-Christina Glock ·
Michael Mayr · Sabrina Luftensteiner
Editors

Database and Expert Systems Applications - DEXA 2024 Workshops

IWCFS, AISys, CIU
Naples, Italy, August 26–28, 2024
Proceedings

Editors
Bernhard Moser
Software Competence Center Hagenberg
Hagenberg, Austria

Lukas Fischer
Software Competence Center Hagenberg
Hagenberg, Austria

Atif Mashkoor
Johannes Kepler University Linz
Linz, Austria

Johannes Sametinger
Johannes Kepler University Linz
Linz, Austria

Anna-Christina Glock
Software Competence Center Hagenberg
Hagenberg, Austria

Michael Mayr
Software Competence Center Hagenberg
Hagenberg, Austria

Sabrina Luftensteiner
Software Competence Center Hagenberg
Hagenberg, Austria

ISSN 1865-0929 ISSN 1865-0937 (electronic)
Communications in Computer and Information Science
ISBN 978-3-031-68301-5 ISBN 978-3-031-68302-2 (eBook)
https://doi.org/10.1007/978-3-031-68302-2

© The Editor(s) (if applicable) and The Author(s), under exclusive license
to Springer Nature Switzerland AG 2024

This work is subject to copyright. All rights are solely and exclusively licensed by the Publisher, whether the whole or part of the material is concerned, specifically the rights of translation, reprinting, reuse of illustrations, recitation, broadcasting, reproduction on microfilms or in any other physical way, and transmission or information storage and retrieval, electronic adaptation, computer software, or by similar or dissimilar methodology now known or hereafter developed.
The use of general descriptive names, registered names, trademarks, service marks, etc. in this publication does not imply, even in the absence of a specific statement, that such names are exempt from the relevant protective laws and regulations and therefore free for general use.
The publisher, the authors and the editors are safe to assume that the advice and information in this book are believed to be true and accurate at the date of publication. Neither the publisher nor the authors or the editors give a warranty, expressed or implied, with respect to the material contained herein or for any errors or omissions that may have been made. The publisher remains neutral with regard to jurisdictional claims in published maps and institutional affiliations.

This Springer imprint is published by the registered company Springer Nature Switzerland AG
The registered company address is: Gewerbestrasse 11, 6330 Cham, Switzerland

If disposing of this product, please recycle the paper.

Preface

Welcome to the Proceedings of DEXA 2024 Workshops. This year, we hosted three workshops: the 9th International Workshop on Cyber-Security and Functional Safety in Cyber-Physical Systems (IWCFS 2024), the 4th International Workshop on AI System Engineering: Math, Modelling and Software (AISys 2024), and the 2nd International Workshop on Certainty in Uncertainty: Exploring Probabilistic Approaches in AI (CIU 2024). These events took place from August 26–28, 2024, in Naples, Italy.

This compilation of papers and presentations represents a convergence of cutting-edge research, interdisciplinary collaboration, and innovative solutions at the forefront of technology and science.

In an era where digital transformation is reshaping industries, the importance of robust cyber-security and seamless integration of functional safety in cyber-physical systems cannot be overstated. These systems, ranging from industrial automation to autonomous vehicles, require rigorous safety protocols and resilient security measures to combat an increasingly sophisticated landscape of threats. Contributions to IWCFS 2024 explored novel methodologies, frameworks, and practical implementations aimed at enhancing the safety and security of cyber-physical systems.

CIU 2024 delved into one of the most intriguing aspects of artificial intelligence research. As AI systems grow more complex and integral to decision-making processes, understanding and managing uncertainty becomes crucial. This workshop highlighted advancements in probabilistic models, Bayesian inference, and uncertainty quantification techniques. These approaches not only enhance the robustness and reliability of AI systems but also pave the way for new applications where traditional deterministic models fall short.

AISys 2024 focused on the foundational aspects of developing AI systems. This workshop emphasized the critical role of mathematical rigor, precise modelling, and sophisticated software engineering in creating powerful and dependable AI solutions. Papers in this workshop addressed a wide range of topics, from algorithmic innovations and computational frameworks to the practical challenges of deploying AI in real-world environments.

The three workshops attracted 24 submissions, from which only 10 papers were accepted after a rigorous peer-review process, ensuring the highest standards of quality and relevance to the DEXA theme.

We extend our deepest gratitude to the authors, reviewers, and organizers whose dedication and expertise made these workshops possible. Their collective efforts ensured that the knowledge shared here is of the highest quality and relevance.

As you engage with these proceedings, we hope you find inspiration, new ideas, and valuable insights that will further your work and contribute to the ongoing evolution of these critical fields.

August 2024

Bernhard Moser
Lukas Fischer
Atif Mashkoor
Johannes Sametinger
Anna-Christina Glock
Michael Mayr
Sabrina Luftensteiner

Organization

Steering Committee

Gabriele Kotsis	Johannes Kepler University Linz, Austria
A Min Tjoa	Vienna University of Technology, Austria
Lukas Fischer	Software Competence Center Hagenberg, Austria
Bernhard Moser	Software Competence Center Hagenberg, Austria
Ismail Khalil	Johannes Kepler University Linz, Austria
Nicola Mazzocca	University of Naples Federico II, Italy
Elio Masciari	University of Naples Federico II, Italy

IWCFS 2024 Chairs

Atif Mashkoor	LIT Secure & Correct Systems Lab, Austria
Johannes Sametinger	Johannes Kepler University Linz, Austria

IWCFS 2024 Program Committee

Paolo Arcaini	National Institute of Informatics, Japan
Irum Inayat	National University of Computers and Emerging Sciences, Pakistan
Jean-Pierre Jacquot	LORIA, Henri Poincaré University, France
Saif Ur Rehman Khan	Shifa Tameer-e-Millat University, Pakistan
Rudolf Ramler	Software Competence Center Hagenberg, Austria
Neeraj Singh	INPT-ENSEEIHT/IRIT, University of Toulouse, France
Michael Vierhauser	University of Innsbruck, Austria
Edgar Weippl	University of Vienna, Austria

AISys 2024 Chairs

Paolo Meloni	University of Cagliari, Italy
Michael Lunglmayr	Johannes Kepler University Linz, Austria

Gerald Czech — Upper Austrian Fire Brigade Association, Austria
Bernhard Moser — Software Competence Center Hagenberg, Austria

AISys 2024 Program Committee

Bernhard Heinzl	Software Competence Center Hagenberg, Austria
Dolly Sapra	University of Amsterdam, Netherlands
Elmar Kiesling	Vienna University of Economics and Business, Austria
Florian Sobiecky	Software Competence Center Hagenberg, Austria
Franz Kraus	University of Mannheim, Germany
Lukas Fischer	Software Competence Center Hagenberg, Austria
Maqbool Khan	Pak-Austria Fachhochschule, Pakistan
Werner Zellinger	Austrian Academy of Sciences, Austria

CIU 2024 Organising Committee

Anna Christina Glock	Software Competence Center Hagenberg, Austria
David Baumgartner	Software Competence Center Hagenberg, Austria
Michael Mayr	Software Competence Center Hagenberg, Austria
Sabrina Luftensteiner	Software Competence Center Hagenberg, Austria

Organizers

Abstracts of Keynote Talks

Multimodal Deep Learning in Medical Imaging

Carlo Sansone

Department of Electrical Engineering and Information Technology, University of Naples Federico II, Italy

Abstract. In this talk, we will consider how Deep Learning (DL) approaches can profitably exploit the presence of multiple data sources in the medical domain.

First, the need to be able to use information from multimodal data sources is addressed. Starting from an analysis of different multimodal data fusion techniques, an innovative approach will be proposed that allows the different modalities to influence each other.

However, in medical applications it is often very difficult to obtain high quality and balanced labelled datasets due to privacy and sharing policy issues. Therefore, several applications have leveraged DL approaches in data augmentation techniques, proposing models that can create new realistic and synthetic samples. Consequently, a new data source can be identified, namely a synthetic data source. In this context, a data augmentation method based on deep learning, specifically designed for the medical domain, will be presented. It exploits the biological characteristics of images by implementing a physiologically aware synthetic image generation process.

Digital Humanism as an Enabler for a Holistic Socio-Technical Approach to the Latest Developments in Computer Science and Artificial Intelligence

A Min Tjoa

TU Wien (Vienna University of Technology), Austria

Abstract. The rapid development of computer science and artificial intelligence (AI) has brought about transformative changes, but not without significant ethical, social, and technical challenges. As early as 2017, Tim Berners-Lee, the inventor of the World Wide Web, warned of the "nasty storm" threatening the future of the web, including the proliferation of fake news, propaganda, and increasing polarization. These issues highlight the urgent need for a paradigm that ensures technology serves the best interests of humanity.

This keynote will explore the foundational principles of Digital Humanism and its role in guiding the development of computer science and AI to align with human values and societal well-being. In December 2023, the United Nations Advisory Panel on AI released its interim report, "Governing AI for Humanity," which highlights the need for AI governance to address challenges and harness AI's potential in an inclusive way, ensuring that no one is left behind. A key measure of AI's success will be its contribution to achieving the SDGs.

The keynote will illustrate how Digital Humanism can be operationalized to create technologies that enhance human capabilities and societal well-being. It will highlight the need for interdisciplinary research and development to harness the potential of computer science and AI for the benefit of humanity.

Digital Humanism offers a vital pathway for navigating the complexities of modern technological advancements. By taking a holistic socio-technical approach, it can be ensured that developments in computer science and AI are aligned with our core human values, thereby fostering a more just, ethical, and sustainable digital future.

Deep Entity Processing in the Era of Large Language Models: Challenges and Opportunities

Toshiyuki Amagasa

University of Tsukuba, Japan

Abstract. Handling entities has long been a critical task in data analytics and integration, with over 80% of time and effort often devoted to data preprocessing. Improving the performance of this task has been a persistent challenge. Recently, transformer-based pre-trained language models and large language models (LLMs) have emerged as key tools for entity processing tasks such as named entity recognition (NER) and entity matching. However, these models introduce new challenges, including significant demands for computational resources and high-quality training data. In this talk, we will review recent advances in deep entity processing and explore the associated challenges and research opportunities.

Contents

Cyber-Security and Functional Safety in Cyber-Physical Systems

Improved CPSoS Security: An Enhanced Anomaly-Based Intrusion
Detection Architecture 3
   *Marco Stadler, Michael Riegler, Johannes Sametinger,
and Christoph Schönegger*

Assessing the Maturity of Blockchain-Based Implementations
with Software Reliability Growth Models 14
   *Muhammad Azeem, Saif Ur Rehman Khan, Atif Mashkoor,
Abdullah Yousafzai, and Habib Un Nisa*

An Automated Ontology-Based Requirements Traceability Technique
in Agile Software Development Context 29
   *Saif Ur Rehman Khan, Uswa Aslam, Atif Mashkoor, Irum Inayat,
and Habib Un Nisa*

Toward a Knowledge-Based Anomaly Identification System for Detecting
Anomalies in the Smart Grid 44
   Sarita Paudel and Abdelkader Magdy Shaaban

AI System Engineering: Math, Modelling and Software

On the Solvability of the XOR Problem by Spiking Neural Networks 57
   Bernhard A. Moser and Michael Lunglmayr

Risk Assessment in AI System Engineering: Experiences and Lessons
Learned from a Practitioner's Perspective 67
   *Magdalena Fuchs, Lukas Fischer, Alessio Montuoro, Mohit Kumar,
and Bernhard A. Moser*

From Paper to Pixels: A Multi-modal Approach to Understand and Digitize
Assembly Drawings for Automated Systems 77
   Raphael Seliger, Sebnem Gül-Ficici, and Ulrich Göhner

Certainty in Uncertainty: Exploring Probabilistic Approaches in AI

Uncertainty Estimation of Raters' Performance and Ground Truth Through
a Bayesian Extension of STAPLE 91
   Davide Cazzorla and Corrado Mencar

Uncertainty Estimation for Energy Consumption Nowcasting 102
   Danel Rey-Arnal, Ibai Laña, and Pablo G. Bringas

Knowledge Guided Clustering Medieval Polychromy 115
   Florian Sobieczky and Elisabeth Sobieczky

Author Index 127

Cyber-Security and Functional Safety in Cyber-Physical Systems

Improved CPSoS Security: An Enhanced Anomaly-Based Intrusion Detection Architecture

Marco Stadler[1]([✉])[iD], Michael Riegler[2][iD], Johannes Sametinger[1][iD],
and Christoph Schönegger[2]

[1] LIT Secure and Correct Systems Lab/Institute of Business Informatics – Software Engineering, Johannes Kepler University, Linz, Austria
{marco.stadler,johannes.sametinger}@jku.at
[2] IT Platforms and Operations, ENGEL AUSTRIA GmbH, Schwertberg, Austria
{michael.riegler,christoph.schonegger}@engel.at
https://www.jku.at/en/lit-secure-and-correct-systems-lab,
https://www.se.jku.at, https://www.engelglobal.com

Abstract. The prevalence of security risks and the subsequent attacks on Cyber-Physical Systems (CPSs) have reached unprecedented levels. Anomaly-based Intrusion Detection Systems (Ab-IDS) emerged in response to this phenomenon, with the purpose of alerting actors in the event of an attack. However, with the advent of Industry 4.0 and smart manufacturing, CPSs have evolved into progressively more complex systems, giving rise to the Cyber-Physical System of Systems (CPSoS) that imposes a wide range of requirements. In light of this problem, we first identify a set of requirements in this problem space and further derive a novel architecture design for that aims to incorporate Ab-IDS into complex CPSoS environments.

Keywords: Intrusion Detection System · System of Systems · Anomaly Detection

1 Introduction

Cyber-Physical Systems (CPSs) integrate computational and physical processes into a single system. Physical processes are typically monitored and controlled by embedded computers and networks via feedback loops in which physical processes influence computations and vice versa. One notable attribute of CPSs that distinguishes these devices from "embedded systems" is their networking capability [8]. This attribute enables CPSs to function as a catalyst for the advent of the fourth industrial revolution and allows for the usage of CPSs in many different domains [6].

With this rise of interconnectivity, CPSs are used within larger ecosystems such as smart energy/water grids, smart cities, or smart manufacturing [5]. The integration of multiple complex (sub-)systems into another higher-level system is referred to as *System of Systems* (SoS) [7]. As a result, CPSs are nowadays regularly managed as so-called *Cyber-physical System of Systems* (CPSoS). From unconnected systems, the CPSoS then interconnects the hardware, software, and even individuals. Subsequently, the data produced by each constituent element may have an influence on other sub-systems, as it will aid in the compilation of comprehensive CPSoS knowledge information [5].

However, knowledge of the interdependent dynamics of computers, software, networks, and physical processes is necessary for the design of such systems [3]. Depending on the environment the systems are deployed and used, special requirements apply [4].

For instance, CPSoS are frequently deployed in *Operational Technology* (OT) environments. OT systems are computing systems designed to manage, control, and monitor industrial operations and tangible equipment. Programmable Logic Controllers, Remote Terminal Units, and Actuators are examples of such systems found in OT infrastructures [12]. With such systems at hand, the International Electrical Commission has established the networks and protocol specifications for factory and process automation. An overall culture of maximizing the longevity, reliability, and availability of investments made to manufacture physical products has developed within OT [4].

This contrasts with the technology used typically in CPSoS. Regular IT structures necessitate immediate responses to emerging requirements [4]. This conflict of interest poses a particular challenge in the context of security. OT systems are concerned with performance, while security engineers prioritize security requirements (e.g., integrity) [22].

For future CPSoS security, therefore, novel approaches and methodologies to reconcile these divergent objectives will be necessary. Intrusion Detection Systems (IDS) play a pivotal role in this domain. Hence, we focus on Anomaly Detection (AD), an aspect of IDS, and propose a method for integrating AD into a CPSoS-enhanced architecture for use, for instance, in OT environments.

This paper is structured as follows: Sect. 2 provides a motivating example, fundamentals are clarified in Sect. 3, followed by an outline of our enhanced architecture in Sect. 4. We discuss our proposed architecture in Sect. 5. Finally, we present related work in Sect. 6, limitations of our approach in Sect. 7, and conclude our findings in Sect. 8.

2 Motivation

SoS (e.g., collaborative systems of production machines, conveyor belts, and edge devices), which are frequently manufactured or assembled by a single company, pose a distinct variety of cybersecurity challenges in the manufacturing domain. These systems are highly customizable to suit individual customer use cases, ranging from operations requiring 24/7 high throughput to those functioning merely a few hours a day, with variations extending to the integration of custom IT systems into the manufacturing process, usage of local networks, or reliance on cloud-based data management. This heterogeneity renders the deployment of off-the-shelf Ab-IDS infeasible, as these solutions typically cannot accommodate the diverse operational profiles nor effectively exchange intelligence within the intricate SoS structure.

Consequently, there is a pressing need for a new IDS architecture that not only dynamically establishes thresholds for AD but is also capable of navigating the complexity of SoS, ensuring robust security tailored to the specific requirements of each manufacturing ecosystem.

3 Background

An IDS is designed to collect and analyze data from multiple sources to detect potential security breaches. Intrusion detection refers to the process of identifying and detecting activities that aim to undermine the confidentiality, integrity, or availability of a given system [13].

IDS can be classified into three distinct categories [9]:

1. **Signature-based**: A signature is a string or pattern that is associated with a recognized threat or attack. Signature-based IDS is the procedure by which documented events are compared to patterns in order to identify potential intrusions.
2. **Anomaly-based**: Anomalies are deviations from known behaviors, whereas profiles depict anticipated or typical behaviors obtained through the long-term monitoring of consistent activities, network connections, hosts, or users. Various attributes, such as the number of e-mails sent, failed login attempts, processor utilization, and so forth, can be utilized to generate these static or dynamic profiles.
3. **Stateful protocol analysis**: The word "stateful" signifies that IDS can discern and track the states of the protocol, such as the pairing of requests and responses.

Although all IDS varieties should be incorporated into a comprehensive security defense strategy, this article focuses specifically on Ab-IDS.

Anomalies are data patterns that deviate significantly from a precisely defined notion of typical behavior. Anomalies may arise in the data due to diverse factors, including malicious activity (e.g., credit card fraud, cyber-intrusion, or system failure). Nevertheless, these anomalies share the characteristic of capturing the interest of the analyst. The practical significance or intrigue of anomalies constitutes a fundamental characteristic of AD [2].

Ab-IDSs use this concept to detect potential threats. The system generates a baseline profile of typical system, network, or program activity before initiating AD. Subsequently, any action that diverges from the established baseline is regarded as a potential intrusion [13]. In a practical context, this might entail an IDS establishing a foundational web socket connection that transmits a heartbeat signal at regular 5-second intervals. The IDS will identify an anomaly if this protocol is subsequently employed to transmit additional (malicious) data in the intervals between these heartbeats, as the baseline will not correspond to the learned baseline.

A typical approach to AD comprises two distinct phases: training and testing. The former involves establishing the typical traffic profile, while the latter involves applying the learned profile to new data [13].

Respectively, to capture the distinct behavior (e.g., network traffic) and accurately identify deviations from the baseline, these two phases must be executed for each system in which AD is implemented. This requires understanding the baseline profile of a programmable logic controller, production machine, or edge device in the context of OT systems. One might intuitively propose establishing a solitary baseline for every system type and subsequently applying the learned model to all analogous systems. However, establishing this baseline for these devices is a complex task within the industry.

For example, the deployment of production machines varies: an "off-the-shelf" baseline cannot be applied to every machine due to factors such as legacy systems, additional hardware components, production cycles, and operating speeds. This issue becomes more complicated when multiple systems operate in conjunction as a SoS, or as components of a larger ecosystem. As a result, we endorse our conceptual approach in this article regarding the efficient implementation of AD in IDS as a component of resilient CPSoS.

4 Enhanced Security Architecture

In this section, we extend the components of a generic IDS architecture with SoS requirements to transform it into a more holistic conceptual architecture.

4.1 Generic Architecture

Figure 1 displays a generalized IDS architecture [1,13]. Solid arrows represent a control and/or data flow, while dashed arrows indicate the response flow to an intrusive activity. First, in the *Audit Collection* phase ①, the data to be analyzed is gathered by the IDS. Sources are manifold and include host/network activity logs, command-based logs, application-based logs, etc. For later reference, the IDS data is stored in an *Audit Storage* ②. The *Analysis and Detection* phase ③ represents the heart of the architecture: The algorithms detect suspicious activities. To accomplish this task, the IDS is required to use two kinds of additional data. *Configuration Data* ④ with information on how, when to collect audit data, and what happens in the event of intrusion. *Reference Data* ⑤ contains the baseline profiles for the AD. Intermediate results (e.g., partially fulfilled intrusions) are cached in the *Active/Processing Data* component ⑥. Output generated by the IDS is handled by the *Alarm* ⑦, where either an automated response ⑧ is triggered or an alert for a security officer ⑨ is triggered.

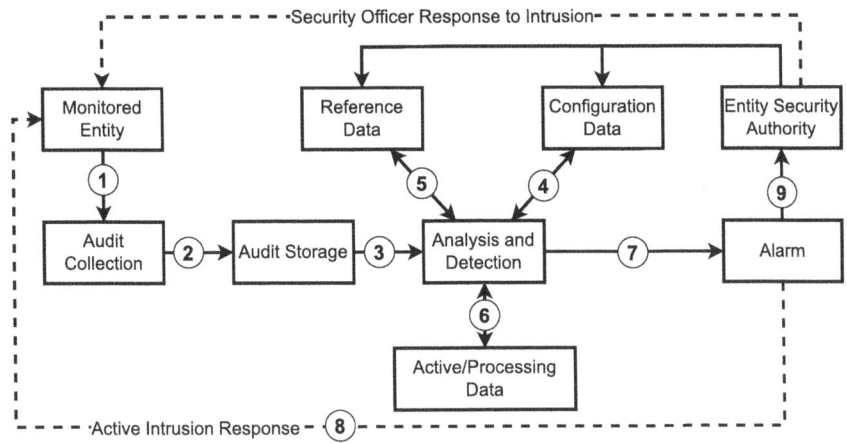

Fig. 1. Generic Intrusion Detection System Architecture [1,13]

4.2 System of Systems Requirements

Although the architecture illustrated in Fig. 1 offers valuable insight for most fundamental scenarios, we contend that it needs to be improved for intricate CPSoS scenarios, where complex systems and a multitude of sub-systems require additional communication flows. We bridge this shortcoming by presenting the implications (requirements) of an industry architecture analysis combined with findings from scientific literature. Therefore, we propose to extend the existing architecture based on the following requirements (Req) and components:

- **Detailed System Profiling (Req1)**
A prerequisite for establishing a unique baseline for each system is the implementation of *System Profiling*. The process of establishing a baseline for the normal or malicious behavior of an entity, such as a group of users, systems, or objects, is referred to as profiling. The profiling process involves the collection of pertinent features and variables that are indicative of or have an impact on a system's normal behavior [14,16].

- **Adaptive Baseline Establishment (Req2)**
The utilization of data acquired during system profiling enables the establishment of dynamic thresholds and criteria that facilitate the detection of anomalies. Establishing a benchmark for normal behavior that can be adapted as the system evolves is the objective [16].

- **Machine Learning and Anomaly Detection Techniques (Req3)**:
The dynamic system architecture necessitates the construction of AD models without labeled reference data; in other words, the system must independently learn normal and abnormal behaviors. As a result, we propose implementing machine learning algorithms, or more specifically, unsupervised machine learning

models. We employ unsupervised learning to identify unlabeled data instances, with clustering emerging as the prevailing learning technique. [11].

- **System of Systems Approach (Req4)**:

For SoS or larger ecosystems, it is crucial to consider the individual systems and their interactions and dependencies. Anomalies arise within systems and in the way they communicate or integrate. Therefore, by actively performing SoS integration, new capabilities are unlocked [10].

- **Continuous Improvement and Validation (Req5)**:

No rigidity characterizes SoS and its sub-systems. Adjustments can be made to machine production cycles, software features are introduced, hardware sensors are implemented, or novel attack strategies are identified, among other possibilities. Continuous evaluation of the baseline and the entire process is imperative in light of the dynamic environment. Hence, to validate and enhance the deployed system, we recommend integrating a feedback loop [18].

4.3 Integrative System of Systems Architecture

Based on the identified requirements, we derive an enhanced IDS architecture. The enhanced architecture is depicted in Fig. 2.

We added five new components (*Operational Collection, Operational Storage, System Profiling, SoS Profile Storage, SoS Alarm Storage* and changed the process of one component (*Reference Data* of the generic architecture to meet all requirements. Table 1 displays a matching between the requirements and the (new) system components.

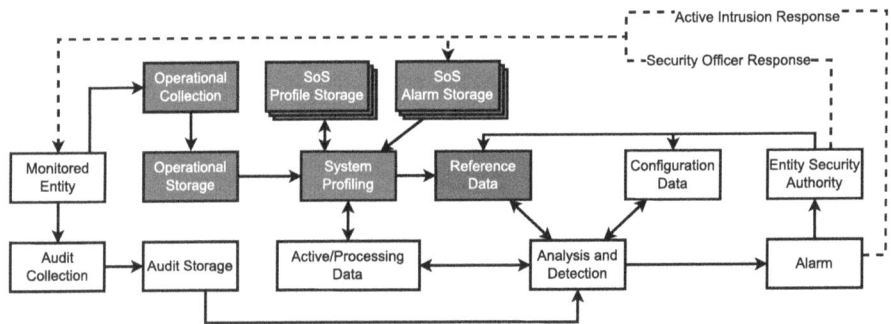

Fig. 2. Enhanced Generic Intrusion Detection System Architecture. *green* = Additions; *blue* = Changes (Color figure online)

Besides collecting regular audit data, the architecture now allows for the collection of operational data (c.f., *Operational Collection*, and *Operational Storage*. This is necessary to derive a comprehensive system profile (c.f., *System Profiling*. Besides the operational data, the profile also uses relevant system profiles provided by other SoS sub-systems (*SoS Profile Storage*. In turn, the generated system profile is then also retrieved by other sub-systems. Using the profiles, machine learning algorithms can now generate an adaptive baseline for the individual system (c.f., *Reference Data*. After deployment of the Ab-IDS, an additional alarm storage component (*SoS Alarm Storage* is activated. This component keeps track of all alarms triggered and provides retrospective incidents to improve the system's profile. The *SoS Alarm Storage* is furthermore the key component to route information back into the IDS process flow. Channeling the response flow (*Security Officer Response/Active Intrusion Response* into the *SoS Alarm Storage* allows for future model adaptations and completes the feedback loop.

Table 1. Component matching with the Requirements

Requirement	Components
Req1	*Operational Collection, Operational Storage, System Profiling*
Req2	*System Profiling*
Req3	*Reference Data*
Req4	*SoS Profile Storage, SoS Alarm Storage*
Req5	*SoS Alarm Storage*

The integration of *Detailed System Profiling* and *Adaptive Baseline Establishment*, complemented by advanced *Machine Learning* algorithms, a *Systems of Systems Approach*, and a *Continuous Improvement and Validation* feedback loop significantly enhances the architecture of the IDS. By leveraging detailed system profiling, the IDS develops an in-depth understanding of normal operational behavior, allowing for the detection of subtle deviations that may indicate advanced threats. Adaptive baseline establishment ensures this understanding dynamically evolves, reducing false positives and negatives and maintaining alignment with the system's changing state. The incorporation of machine learning enables the system to intelligently analyze patterns and predict potential threats based on historical data (i.e., operational data like sensor measurements from the CPS under monitoring and previous alarms), enhancing the accuracy and efficiency of threat detection. Additionally, adopting an SoS approach ensures that the IDS can effectively monitor and protect complex interconnected systems, providing a holistic security stance. The enhanced IDS architecture facilitates continuous improvement and validation through its dynamic adaptation and learning capabilities, ensuring that the system remains effective against evolving threats by constantly refining its detection mechanisms and baselines. This comprehensive, intelligent, and adaptive architecture equips organizations with a robust defense mechanism, capable of countering sophisticated threats in a dynamic digital environment, thereby establishing a superior security posture.

5 Discussion

A significant challenge arises from some customers' reluctance to share operational data among devices, across networks, or with the cloud, which contrasts with the SoS approach integral to our architecture (c.f., *SoS Profile Storage* and *SoS Alarm Storage*). This may result from a company's desire to safeguard business secrets or the personnel responsible for machine operation and maintenance. A potential solution involves the implementation of data localization and anonymization techniques. By ensuring that data can be processed and analyzed locally on devices or within the customer's network-without compromising sensitive information-customers might be more inclined to utilize the SoS capabilities.

Operational data collection (c.f., *Operational Collection*, and *Operational Storage*) poses a formidable challenge within the manufacturing sector. Production machines can generate substantial quantities of data within short periods. As a result, a system must possess not only the capacity to manage such volumes of data but also the proficiency to efficiently process and derive valuable insights from this data. An approach that can be implemented to tackle this challenge involves the integration of edge computing technologies. By implementing data processing on or in close proximity to the devices that generate the data, the system can effectively mitigate latency, improve its ability to analyze data in real time and alleviate the strain on network bandwidth.

Transferring knowledge and expertise to the model via the feedback loop, which entails incorporating human expertise (c.f., *Security Officer Response*) into the architecture, is a challenging task. Implementing a human-in-the-loop approach is of utmost importance when making adjustments to anomaly thresholds and enhancing detection algorithms based on practical observations. To streamline this process, integration can be achieved by developing intuitive, user-friendly interfaces that enable security professionals to effortlessly input and modify data.

6 Related Work

Vieira et al. propose in IDS for grid and cloud computing [21]. The proposed architecture leverages multiple nodes and knowledge transfer between the nodes to share relevant data. While these base principles are similar to our SoS requirements (c.f., **Req4**), their architecture does not focus on gathering operational data to generate an individual IDS model for each system.

Song et al. describe a continuous verification mechanism for SoS [18]. With multiple collaborative feedback loops, they can adapt SoS components at runtime. The described aspects correspond to our feedback loop system components (c.f., **Req5**) but do not focus on IDS or the AD model generation of our architecture.

Some meta-studies analyzed IDS SoS characteristics before. For instance, Zoppi et al. outline a state-of-the-art analysis of Ab-IDS for SoS and derive a set of best practices [23]. It is our firm belief that the findings have the potential to inform the execution of our architecture. However, they lack a specific framework and do not address the issue of establishing a dynamic baseline as an integral component of our problem space.

Concerning our prior research [20], we contend that the enhanced IDS architecture outlined herein enhances our multi-level resilient edge computing system architecture. We describe how multiple feedback loops can be implemented at various computing levels of an SoS architecture to adapt to security incidents at runtime. By combining the enhanced IDS architecture described in this paper with the multi-level feedback loop mechanism, an even more potent set of tools is produced to fortify systems against cyber attacks.

While there is a plethora of related work in the domains of SoS, IDS, and AD, none comprehensively covers the entire process and the improvement aspects in the manner we have outlined. Our approach uniquely integrates detailed system profiling, adaptive baselines, machine learning, SoS methodology, and a continuous validation workflow to enhance detection capabilities and system adaptability in a way not previously described in the literature.

7 Limitations

Like any other approach, our enhanced Ab-IDS concept is subject to limitations.

First of all, depending on the domain in which the CPSoS is deployed, data privacy may play a pivotal role due to obligations enforced by governments/regulatory bodies [15] or because of industry standards (e.g., medical domain [17]). These regulations conflict with the nature of the enhanced IDS architecture. The process relies on the ability of the individual sub-systems to share knowledge within and across the SoS.

Furthermore, the system might be unable to distinguish between alarms triggered because of security incidents or triggers from abnormal system behavior. This could lead to false positives and result in unintended system responses. However, we are convinced that the human-in-the-loop (i.e., the *Entity Security Authority*) can make a huge difference. Educating the person responsible for triggering alarms can help reduce the amount. Furthermore, depending on the application scenario, this could also be a positive use case. Consider a robotic arm working close to humans and acting abnormally. In this case, triggering an alarm that shuts down the robotic arm is essential to ensure safety. It is not important whether the alarm is triggered because of a security incident or because of a misconfiguration during maintenance [19].

8 Conclusion

This work introduces an improved Ab-IDS architecture that can effectively address the varied demands within the CPSoS sector. In this paper, we present a motivational scenario to define the problem space and initially explain a generic architecture from existing literature. We then identify requirements specific to the systems under consideration and propose an improved version of the architecture. We demonstrate that the new architecture includes all the requirements that have been previously identified.

While the architecture we have provided is still in the conceptual stage, we believe that our architecture, requirements, and findings can assist both researchers and practitioners in constructing and identifying system components for implementing Ab-IDS. This applies not only to CPSoS but also to other types of SoS.

We are currently in the process of implementing the architecture described in Section motivation to provide empirical evidence of its practicality and validate it's feasibility.

Acknowledgements. This work has partially been supported by the LIT Secure and Correct Systems Lab funded by the State of Upper Austria.

References

1. Axelsson, S.: Research in intrusion-detection systems: a survey. Technical report, Technical report 98–17. Department of Computer Engineering, Chalmers University of Technology (1998)
2. Chandola, V., Banerjee, A., Kumar, V.: Anomaly detection: a survey. ACM Comput. Surv. **41**(3), 15:1-15:18 (2009). https://doi.org/10.1145/1541880.1541882
3. Derler, P., Lee, E.A., Sangiovanni Vincentelli, A.: Modeling cyber-physical systems. Proc. IEEE **100**(1), 13–28 (2012). https://doi.org/10.1109/JPROC.2011.2160929
4. Felser, M., Rentschler, M., Kleineberg, O.: Coexistence standardization of operation technology and information technology. Proc. IEEE **107**(6), 962–976 (2019). https://doi.org/10.1109/JPROC.2019.2901314
5. Ferrer, B.R., et al.: Towards the adoption of cyber-physical systems of systems paradigm in smart manufacturing environments. In: 2018 IEEE 16th International Conference on Industrial Informatics (INDIN), pp. 792–799 (2018). https://doi.org/10.1109/INDIN.2018.8472061
6. Jazdi, N.: Cyber physical systems in the context of Industry 4.0. In: 2014 IEEE International Conference on Automation, Quality and Testing, Robotics, pp. 1–4 (2014). https://doi.org/10.1109/AQTR.2014.6857843
7. Keating, C., et al.: System of systems engineering. Eng. Manag. J. **15**(3), 36–45 (2003). https://doi.org/10.1080/10429247.2003.11415214
8. Lee, E.A.: Cyber physical systems: design challenges. In: 2008 11th IEEE International Symposium on Object and Component-Oriented Real-Time Distributed Computing (ISORC), pp. 363–369 (2008). https://doi.org/10.1109/ISORC.2008.25
9. Liao, H.J., Richard Lin, C.H., Lin, Y.C., Tung, K.Y.: Intrusion detection system: a comprehensive review. J. Netw. Comput. Appl. **36**(1), 16–24 (2013). https://doi.org/10.1016/j.jnca.2012.09.004
10. Madni, A.M., Sievers, M.: System of systems integration: key considerations and challenges. Syst. Eng. **17**(3), 330–347 (2014). https://doi.org/10.1002/sys.21272
11. Maseer, Z.K., Yusof, R., Bahaman, N., Mostafa, S.A., Foozy, C.F.M.: Benchmarking of machine learning for anomaly based intrusion detection systems in the cicids2017 dataset. IEEE Access **9**, 22351–22370 (2021). https://doi.org/10.1109/ACCESS.2021.3056614

12. Nafees, M.N., Saxena, N., Cardenas, A., Grijalva, S., Burnap, P.: Smart grid cyber-physical situational awareness of complex operational technology attacks: a review. ACM Comput. Surv. **55**(10), 215:1-215:36 (2023). https://doi.org/10.1145/3565570
13. Patcha, A., Park, J.M.: An overview of anomaly detection techniques: existing solutions and latest technological trends. Comput. Netw. **51**(12), 3448–3470 (2007). https://doi.org/10.1016/j.comnet.2007.02.001
14. Peng, T., Leckie, C., Ramamohanarao, K.: Survey of network-based defense mechanisms countering the DoS and DDoS problems. ACM Comput. Surv. **39**(1), 3–es (2007). https://doi.org/10.1145/1216370.1216373
15. Perera, C., Ranjan, R., Wang, L., Khan, S.U., Zomaya, A.Y.: Big data privacy in the internet of things era. IT Professional **17**(3), 32–39 (2015). https://doi.org/10.1109/MITP.2015.34
16. Resende, P.A.A., Drummond, A.C.: Adaptive anomaly-based intrusion detection system using genetic algorithm and profiling. Secur. Privacy **1**(4), e36 (2018). https://doi.org/10.1002/spy2.36
17. Riegler, M., Sametinger, J., Rozenblit, J.W.: Context-aware security modes for medical devices. In: 2022 Annual Modeling and Simulation Conference (ANNSIM), pp. 372–382 (2022). https://doi.org/10.23919/ANNSIM55834.2022.9859283
18. Song, J., Kang, J., Hyun, S., Jee, E., Bae, D.H.: Continuous verification of system of systems with collaborative MAPE-K pattern and probability model slicing. Inf. Softw. Technol. **147**, 106904 (2022). https://doi.org/10.1016/j.infsof.2022.106904
19. Stadler, M., Riegler, M., Sametinger, J.: Towards increasing safety in collaborative CPS environments. In: Database and Expert Systems Applications - DEXA 2023. CCIS, vol. 1872, pp. 79–85. Springer, Cham (2023). https://doi.org/10.1007/978-3-031-39689-2_8
20. Stadler, M., Sametinger, J., Riegler, M.: Cyber-resilient edge computing: a holistic approach with multi-level MAPE-K loops. In: 2024 IEEE 21th International Conference on Software Architecture Companion Proceedings (2024). Accepted for Publication
21. Vieira, K., Schulter, A., Westphall, C.B., Westphall, C.M.: Intrusion detection for grid and cloud computing. IT Professional **12**(4), 38–43 (2010). https://doi.org/10.1109/MITP.2009.89
22. Zhang, X., Wang, X.: Tradeoff analysis for conflicting software non-functional requirements. IEEE Access **7**, 156463–156475 (2019). https://doi.org/10.1109/ACCESS.2019.2949218
23. Zoppi, T., Ceccarelli, A., Bondavalli, A.: Exploring anomaly detection in systems of systems. In: Proceedings of the Symposium on Applied Computing. SAC '17, New York, NY, USA, pp. 1139–1146. Association for Computing Machinery (2017). https://doi.org/10.1145/3019612.3019765

Assessing the Maturity of Blockchain-Based Implementations with Software Reliability Growth Models

Muhammad Azeem[1], Saif Ur Rehman Khan[2(✉)], Atif Mashkoor[3], Abdullah Yousafzai[4], and Habib Un Nisa[2]

[1] Department of Computer Science, COMSATS University Islamabad (CUI), Islamabad, Pakistan
[2] Department of Software Engineering, Shifa Tameer-e-Millat University (STMU), Islamabad, Pakistan
{saif_rehman.ssc,habibunnisa.ssc}@stmu.edu.pk
[3] Institute for Software Systems Engineering, Johannes Kepler University (JKU), Linz, Austria
atif.mashkoor@jku.at
[4] Department of Computer Science, University of Central Punjab (UCP), Lahore, Pakistan
abdullah.yousafzai@ucp.edu.pk

Abstract. Blockchain technology is widely used in healthcare, IoT, smart grids, autonomous vehicles, and so on to improve security, trust, transparency, and reliability in handling data. However, Blockchain-Based Implementations (BBIs) remain a complex and challenging task for the developers, especially when ensuring quality across multiple nodes. Despite the growing popularity of BBIs, the existing literature lacks a way to establish quality standards for blockchain development. To tackle this issue, we propose a conceptual model that uses Software Reliability Growth Models (SRGMs) to assess the code-based maturity of BBIs. The proposed model analyzes reports from different software releases to compare current versions with previous ones. We evaluate well-known BBI platforms, such as Ethereum and Hyperledger Fabric, using bug reports collected from the tracking management system. By applying SRGMs to these reports, we identify factors like fault propagation and remaining bugs, helping to predict testing needs for future releases and identifying areas potentially affected by bugs. We assess software reliability based on bug resolution parameters and establish software maturity metrics to evaluate release completeness by reporting remaining bugs in current and upcoming releases. The results reveal that the proposed conceptual model is effective in empirically assessing the reliability of BBIs.

The work of Atif Mashkoor is supported by the Austrian Science Fund (FWF) grant # I 4744-N and the LIT Secure and Correct Systems Lab sponsored by the province of Upper Austria.

© The Author(s), under exclusive license to Springer Nature Switzerland AG 2024
B. Moser et al. (Eds.): DEXA 2024 Workshops, CCIS 2169, pp. 14–28, 2024.
https://doi.org/10.1007/978-3-031-68302-2_2

Keywords: Blockchain · Ethereum · Hyperledger Fabric · Software Reliability Growth Model · Maturity · Conceptual Model

1 Introduction

Software reliability is crucial to software quality, particularly from a customer-oriented perspective [1]. Over the years, Software Reliability Growth Models (SRGMs) have been developed to quantify software reliability, often utilizing the Non-Homogeneous Poisson Process (NHPP) [2]. However, discrepancies between testing and live environments-such as coverage, effort, fault detection, and correction times-pose challenges. Some SRGMs make unrealistic assumptions, such as no difference between testing and live environments, leading to reliability issues [3]. Despite these challenges, SRGMs are essential for ensuring customer satisfaction [4].

Emerging technologies like blockchain enhance software system reliability and trust. Blockchain's decentralized and secure nature has revolutionized various industries, categorized into permissionless and permissioned blockchains based on user permissions. Consensus mechanisms and smart contracts validate transactions without intermediaries [5].

In Blockchain-Based Industries (BBIs), platforms like Ethereum and Hyperledger Fabric are widely used. With its decentralized approach and support for smart contracts, Ethereum is popular for decentralized applications (DApps) such as DeFi and NFTs [6]. Hyperledger Fabric, with its modular architecture and privacy features, caters to enterprise needs in various industrial applications [7]. Despite challenges in ensuring software reliability, SRGMs, and blockchain technologies offer promising solutions. Motivated by this, we propose a conceptual model leveraging SRGMs to assess the code-based maturity of BBIs. The main research contributions of this work are summarized as follows:

- **RC-1:** Conducting an extensive review and analysis of existing literature in the BBIs context.
- **RC-2:** Empirically evaluating the SRGMs for analyzing the stochastic behavior of detecting and resolving bugs in BBIs. A new class of SRGMs with its fitness techniques is considered for residual bug detection and correction.
- **RC-3:** Assessing the maturity of employed BBIs using the existing SRGMs. To accomplish this, we extract the bug reports and resolution time from the source code available on the GitHub repositories.

The remaining sections of this article are organized as follows: Sect. 2 provides the related work, and Sect. 3 provides the background of software reliability growth models. Section 4 presents the proposed conceptual model, and Sect. 5 mentions working illustrations of the proposed model. Section 6 provides the experimental results, and Sect. 7 reports on the discussion. Finally, Sect. 8 concludes this work.

2 Related Work

The software industry is crucial for advancing sectors such as medicine, sports, travel, and food, with software reliability being essential for ensuring functionality. Research on Software Reliability Growth Models (SRGMs) began in 1971 with Jelinski and Moranda's pioneering work [8], leading to several models, notably S-shaped and concave models.

Early models assumed constant error rates, but error rates vary over time. Non-homogeneous models are now favored in industry and public sector software development. Li and Pham [9] utilized the NHPP model to predict uncertainty in software environments, offering flexibility in modeling random effects. While probability theory has been proposed for reliability [10], software failure differs from hardware failure, as it is more intellectually driven. Hence, probability theory alone may not suffice due to the diverse nature of software and the lack of standardization in the software development life cycle.

Liu et al. [11] introduced a software belief reliability growth model based on probability theory, incorporating parameter estimations. An alternative to probability theory is uncertainty theory, which includes axioms like normality and duality. Belief theory, combining probability and uncertainty, addresses both aleatory and epistemic uncertainties, though these theories are not uniformly applied across all metrics. Hou et al. [12] proposed a framework for NHPP and Non-NHPP SRGMs, categorizing data into examined and unexamined groups. However, this model required extensive computations, potentially burdening code execution in real projects. Another approach integrated SRGMs with deep learning, using feedback algorithms to optimize SRGM parameters and assess performance with real-time datasets [13].

Li et al. [14] proposed a quality control model starting from the software requirement phase, employing SRGMs for reliability measurement and enhancing testing efficiency. Garg et al. [15] developed a CODAS-E framework for selecting and ranking SRGMs based on multiple parameters, highlighting the importance of choosing appropriate SRGMs for specific domains.

Blockchain technology presents unique challenges in software reliability, particularly for SRGMs. With their complex, decentralized architectures, blockchain systems require robust security mechanisms, complicating traditional SRGM applications. Decentralized applications operate across multiple nodes without a central authority, adding complexity to the software architecture and making traditional SRGMs less effective.

Blockchain applications often use smart contracts-self-executing contracts written into code-which can lead to unpredictable reliability growth patterns due to their dynamic nature. Managing the reliability of these dynamically updated contracts is challenging [18]. Integrating blockchain technology with existing systems can introduce compatibility and reliability issues, as discussed by Politou et al. [20] and Bao et al. [21].

The rapidly evolving blockchain landscape, with new standards and protocols, leads to uncertainties and variations in reliability growth. Nguyen et al. [24] highlighted challenges in blockchain cloud applications but did not address how evolving standards impact reliability. SRGMs must adapt to account for these rapid changes in the blockchain ecosystem.

Although many SRGMs have performed well in various domains, none have specifically focused on the maturity and reliability of blockchain-based implementations. All reliability growth models aim to assess the maturity of future software releases, but there is a gap in models addressing blockchain-based implementations.

3 Background of SRGMs

This section describes the core concept of SRGMs and a particular class of models based on the Non-Homogeneous Poisson Process (NHPP) for fault detection and resolution. The NHPP has been very successful and famous for modeling the behavior of large-size open-source projects. Bug detection models are presented in Sect. 3.1, widely used in the software reliability community. Section 3.2 describes the bug resolution process to extend the SRGMs.

3.1 NHPP Bug Detection

For mathematical formulations, assume the software contains bugs (faults) before the testing phase, and the number is a random variable N_0 following the Poisson distribution with a mean α as shown in Eq. 1 [23]. Table 1 shows the elements used in equations.

$$P(N_0 = n) = \frac{\alpha^n}{n!} e^{-\alpha} \tag{1}$$

The cumulative distribution function $F_d(t)$ represents the single bug detection probability over time t. The number of detected bugs $N_d(t)$ with respect to time t is described in Eq. 2 [23]:

$$P(N_d(t) = k | N_0 = n) = \binom{n}{k} F_d(t)^k (1 - F_d(t))^{n-k} \tag{2}$$

Equation 3 [23] describes the probability of observing exactly k bugs by time t.

$$P(N_d(t) = k) = \sum_{n=k}^{\infty} P(N_d(t) = k | N_0 = n) = \frac{[\alpha F_d(t)]^k}{k!} e^{-\alpha F_d(t)} \tag{3}$$

Many software reliability features can be estimated from fault detection's mean value function $m(t)$. The expected number of detected bugs by time t is given by Eq. 4 [23]:

$$E[N_d(t)] = m(t) = \alpha F_d(t) \tag{4}$$

The instantaneous bug detection rate can be computed by Eq. 5 [23]:

$$\lambda(t) = \frac{dm(t)}{dt} = \alpha f_d(t) \tag{5}$$

Suppose the number of initial faults in the blockchain software is finite $\lim_{t\to\infty} m(t)$. Thus, the residual (undetected) faults are defined as in Eq. 6 [23]:

$$r(t) = E[\alpha - N_d(t)] = \alpha - m(t) \tag{6}$$

The software reliability of detecting a new bug in the time interval $(t, t + x]$ is known as a conditional probability as described in Eq. 7 [23]:

$$R(x|t) = e^{-\int_t^{t+x} \lambda(x)dx} = e^{Rdm(t)-Rdm(x+t)} \tag{7}$$

There are two types of cost for software: (i) the cost of testing $c_t(t)$ before release, and (ii) the cost of removing bug $c_w(t)$ from the software in the operational phase during the warranty period T_w of the software life cycle. The total cost of software maintenance by assuming that software is released after the testing time T is represented in Eq. 8 [23]:

$$RdC(T) = \int_{t=0}^{T} Rdc_t(t)dt + \int_{t=T}^{T+T_w} Rdc_w(t)\lambda(t)dt \tag{8}$$

For fault detection purposes, we have selected eight widely used NHPP models, including Musa-Logarithmic (MUSA log), Goel-Okumoto Exponential (GO Exp), Generalized Goel-Okumoto (GGO), Inflection S-shaped (ISS), Delayed S-Shaped (DSS), Yamada-Exponential (YEX), Gompertz (GOMP) and Logistics (LOGIST). The considered NHPP models are already used to assess the reliability of components of SRGM-based controller software. A summary of the models is given in Table 2, in which mean value function and failure intensity are available. The characteristics of the considered NHPP models are that three out of eight are concave, and the rest of the five are S-shaped curved models. Moreover, seven are finite failure detection, and one is an infinite failure detection model.

3.2 Fault Resolution

The bug resolution process contains two steps: (i) bug detection and (ii) resolution. In this work, we assume that bug resolution steps are independent. The densities of bug detection and resolution are described in Eq. 9 [25]:

$$f_r(t) = \int_{x=0}^{t} f_d(t-x)f_c(x)dx = [f_d * f_c](t) \tag{9}$$

where $f_r(t)$ represents the density of bug detection and $f_c(t)$ represents bug resolution density. The mean value function of bug resolution is described in Eq. 10 [25]:

$$m_r(t) = \alpha F_r(t) = \alpha \int_{\tau=0}^{t} [f_d * f_c](\tau)d\tau \tag{10}$$

Equation 11 [25] is used for the fault resolution process during SRGMs generation from arbitrary distributions. Due to both bug detection and resolution being

Table 1. Symbols with explanation

Symbol	Explanation
N	Poisson distribution
α	Mean value
n	Number of bugs
$F_d(t)$	Arbitrary distribution
k	Constant
N_0	Random variable
N_d	Number of detected bugs
t	Time to detect
$m(t)$	Mean value of time
$\lambda(t)$	Instantaneous bug detection
$r(t)$	Residual bugs arrival time
E	Estimated bugs
n	Reliability coefficient
Rdf_d	Residual bug detection
Rdf_c	Residual bug correction
$R(x\|t)$	Conditional software reliability

Table 2. Fault Detection Approaches of Non-Homogeneous Poisson Process

Model	Abbreviation	Shape	Mean value function	Failure intensity
Musa-Okumoto logarithmic	MUSA(Log)	Concave	$m_{mo}(t) = \alpha \ln(1+bt)$	$\lambda_{\log}(t) = \frac{ab}{1+bt}$
Goel-Okumoto exponential	GO(Exp)	Concave	$m_{go}(t) = \alpha(1-e^{-bt})$	$\lambda_{go}(t) = abe^{-bt}$
Generalized Goel-Okumoto	GGO	S-Shaped	$m_{ggo}(t) = \alpha(1-e^{-bt^c})$	$\lambda_{ggo}(t) = abct^{c-1}e^{-bt^c}$
Ohba's inflection S-shaped	ISS	S-Shaped	$m_{iss}(t) = \alpha \frac{1-e^{-bt}}{1+\phi e^{-bt}}$	$\lambda_{iss}(t) = abe^{-bt}\frac{1+\phi}{(1+\phi e^{-bt})^2}$
Yamada delayed S-shaped	DSS	S-Shaped	$m_{dss}(t) = \alpha(1-(1+bt)e^{-bt})$	$\lambda_{dss}(t) = ab^2 te^{-bt}$
Yamada exponential	YEX	Concave	$m_{yex}(t) = \alpha(1-e^{-r(1-e^{-bt})})$	$\lambda_{yex}(t) = abre^{-bt}e^{-r(1-e^{-bt})}$
Gompertz	GOMP	S-Shaped	$m_{gomp}(t) = \alpha k^{b^t}$	$\lambda_{gomp}(t) = \alpha \ln b \ln k b^t k^{b^t}$
Logistic	LOGIST	S-Shaped	$m_{logist}(t) = \frac{\alpha}{1+ke^{-bt}}$	$\lambda_{logist}(t) = \frac{abke^{-bt}}{(1+ke^{-bt})^2}$

Goel-Okumoto processes, there is an integral-based close-form solution for the proposed model with limited combinations.

$$m_r^{go-go}(t) = \alpha[1 - \frac{Rd_1 e^{-Rd_2 t} - Rd_2 e^{-Rd_1 t}}{Rd_1 - Rd_2}] \quad (11)$$

Putting the integral values in Eq. 11 with its Piecewise Constant Approximation (PCA), a numerical approximation of NHPP for an arbitrary combination can be obtained and used to fit fault report data, as mathematically modeled in Eq. 12 [25]:

$$F_r(t) = \lim_{\triangle x \to 0} \sum_{i=0}^{n=t/\triangle x} [Rdf_d * Rdf_c](i\triangle x)\triangle x \quad (12)$$

3.3 Model Parameter Fitting

We employed the Least Square Estimation (LSE) method to fit model parameters, minimizing the squared distance between observed and expected data. The Levenberg-Marquard algorithm and the Trusted Region Reflective (TRF) algorithm were used to resolve uncertainties in model selection and fit the parameters within specified bounds.

The Goodness of Fit (GOF) was measured using Mean Square Error (MSE), Theil's Statistics (TS), and Coefficient of Determination (R^2). These metrics are defined in Eqs. 13, 14, and 15 [25]:

$$MSE = \frac{1}{k}\sum_{i=1}^{k}(Rdm(t_i) - Rdm_{est}(t_i))^2 \tag{13}$$

$$TS = \sqrt{\frac{\sum_{i=1}^{k}(Rdm(t_i) - Rdm_{est}(t_i))^2}{\sum_{i=1}^{k}Rdm(t_i)^2}} * 100\% \tag{14}$$

$$R^2 = 1 - \frac{\sum_{i=1}^{k}(Rdm(t_i) - Rdm_{est}(t_i))^2}{\sum_{i=1}^{k}(Rdm(t_i) - \bar{Rdm})^2} \tag{15}$$

where $m(t_i$ is observed data, m_{est} is the estimated data by model, at time t of ith fault report and $\bar{Rdm} = \frac{1}{k}\sum_{i=1}^{k}(Rdm(t_i))$.

4 Proposed Conceptual Model

In this section, we proposed a conceptual model for evaluating the maturity of Blockchain-Based Implementations (BBIs). The proposed model contains four main steps, as depicted in Fig. 1.

4.1 Data Collection

The first step involves gathering comprehensive bug report data. This includes documenting all detected bugs and detailing the efforts to resolve them, including time and resources spent. This thorough data collection is essential for the subsequent steps.

4.2 Model Selection

After data collection, appropriate Software Reliability Growth Models (SRGMs) are selected for bug detection and resolution. Models are evaluated for performance, and parameters are fine-tuned using methods like Non-Homogeneous Poisson Process (NHPP) and Least Square Estimation (LSE).

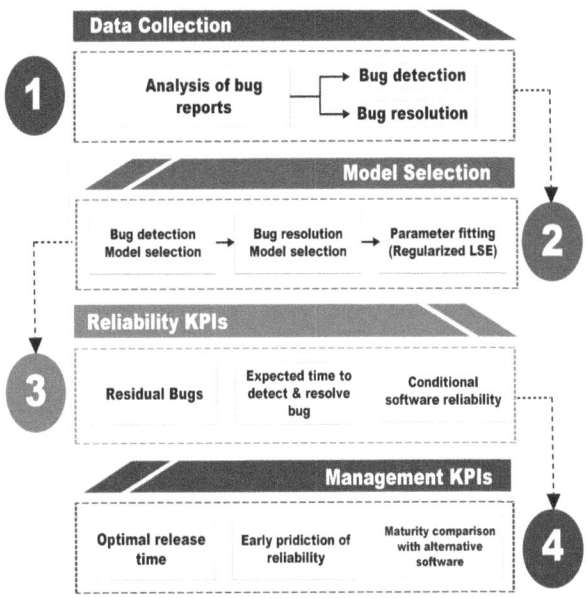

Fig. 1. Overview of the Proposed Conceptual Model

4.3 Reliability KPIs

Reliability KPIs are identified to assess SRGMs, including tracking residual bugs, estimating detection and resolution times, and evaluating software reliability under specific conditions to enhance future reliability.

4.4 Management KPIs

Management KPIs focus on strategic decisions such as optimal release time, predicting future reliability, and benchmarking against similar products to identify improvement areas.

5 Working Illustration of the Proposed Conceptual Model

This section provides an overview and comparison of bug management systems and release cycles for two prominent blockchain-based implementation (BBI) platforms: (i) Ethereum and (ii) Hyperledger Fabric.

5.1 Ethereum

GO Ethereum focuses on providing blockchain access, enabling various operations like creating accounts, transferring funds, deploying smart contracts, and interacting with the Ethereum network. It's utilized by over 200 organizations, including Samsung Group, J.P. Morgan, Mastercard, and Microsoft, with over 100 developers contributing. GO Ethereum follows a monthly release schedule, adopting an incremental development culture with two-week sprints. Since its inception in July 2014, Ethereum has completed 163 releases. Bug reports for Ethereum are managed via ZenHub, an external GitHub extension focusing on issues labeled as bugs. Due to insufficient bug reports in the latest release, Nemata, statistical data analysis is conducted on previous versions, Jarfor and Pongea Expanse.

5.2 Hyper-Ledger Fabric

It is an enterprise-grade permissioned distributed ledger framework for solutions and applications featuring a modular-based and versatile design. With over 100 developers contributing, Hyperledger Fabric has 231,627 lines of code. Unlike Ethereum, it follows an irregular release cycle, ranging from one to six months, with 50 releases completed since its first release on Nov 22, 2019. Hyperledger Fabric adopts an agile methodology similar to Ethereum but is managed using the Jira management tracking system. Jira provides comprehensive bug tracking information, including bug-affected areas, priority, description, reproducing steps, creation date, and resolution date. Reports are classified into release and severity categories, focusing solely on bug-related issues.

6 Experimental Results

The experimental setup involves implementing the mathematical modeling in C# language using the Scikit library, with graphical results plotted using the Matlab library Matplotlib. The experiments were designed to analyze bug reports, apply SRGMs to detect and categorize bugs and evaluate bug resolution times using SRGMs. The overall goal was to determine the best SRGM for residual bug detection and measure the goodness-of-fit (GOF) of the proposed models.

The first step involved organizing bug reports from Ethereum and Hyperledger Fabric. These reports were processed using various SRGMs to categorize bugs. Figure 2 show the heatmaps of bug detection by different SRGMs in Ethereum.

Following categorization, we used these SRGMs to predict bug resolution times. The Least Squares Error (LSE) for different bug types was calculated, and the best SRGM was determined based on bug detection and resolution performance. The GOF of the models was measured using Mean Square Error (MSE), Theil's Statistics (TS), and the Coefficient of Determination (R^2). MSE was used for individual releases, TS for comparison across all releases, and R^2 to measure data variance explained by the model.

The release-wise bug detection and resolution times for Ethereum are shown in Fig. 3. Figure 3 shows the number of detected and resolved bugs across various

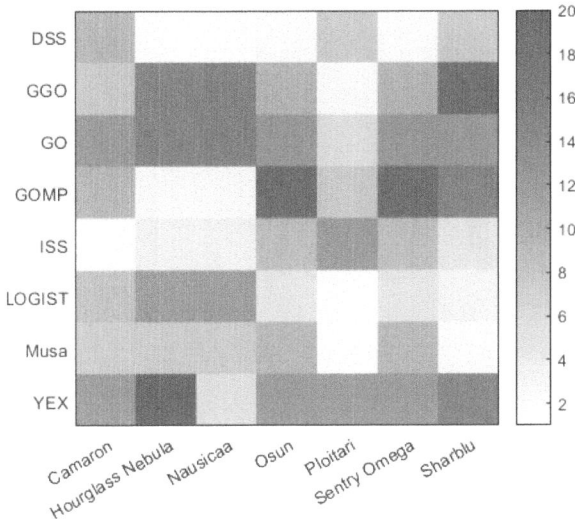

Fig. 2. Previous SRGM Models Detecting Bugs in Ethereum

Ethereum releases from December 2021 to August 2022. The y-axis represents the number of bugs, with detected bugs ranging from 100 to 150 and resolved bugs from 70 to 90. Each release, marked on the x-axis, displays continuous fluctuations in both detection and resolution, indicating an ongoing process. Bug detection consistently exceeds bug resolution, highlighting the perpetual nature of managing software reliability where new bugs are continuously identified and resolved over successive releases.

The Goodness-of-Fit (GOF) is evaluated and illustrated in Fig. 4. Previous releases exhibited a significant number of detected and resolved bugs. We focused on the most recent releases to assess the median bug detection and resolution times. Bug detection in prior releases often occurred quicker than expected, while resolution times exceeded expectations.

In previous releases, bug resolution typically took longer than bug detection. For instance, in the release of Nausicaa, the median bug detection time was 40 h, while bug resolution time approached 200 h. This disparity was attributed to UI/UX bugs that required less reproduction time but longer resolution efforts. Conversely, in the release of Sentry Omega, a significant gap between bug detection and resolution times was observed due to complexities in bug reproduction. Certain bugs necessitated challenging scenarios to recreate in testing and UAT environments, leading to prolonged detection, reproduction, and resolution times. Some data points fell outside the quartile ranges, denoted by a (+) sign, signifying outliers.

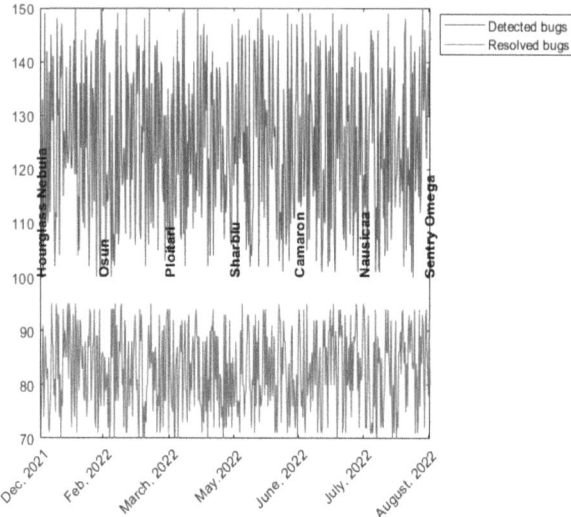

Fig. 3. Release-Wise Bug Detection and Resolution of Ethereum

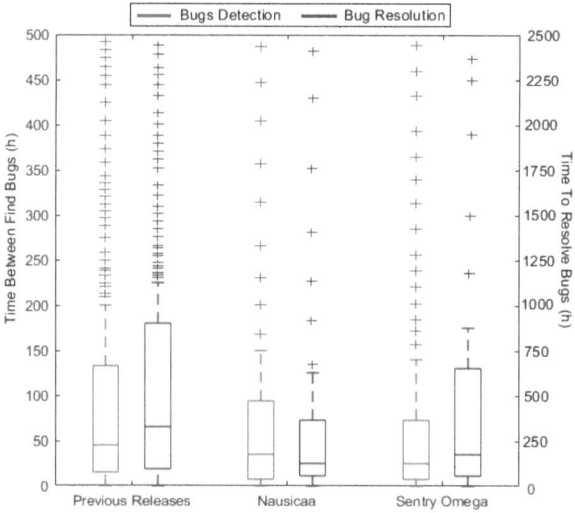

Fig. 4. Median of Bug Detection and Resolution in Ethereum

Hyperledger Fabric's bug detection and resolution trends, depicted in Fig. 5, reflect varying median times. Earlier releases showed median bug detection times of 25 h, contrasting with median bug resolution times nearing 200 h. This disparity was primarily due to data correction and bug reproduction complexities, with residual bugs from previous releases also impacting subsequent releases and contributing to extended resolution times. The comparison of the proposed model

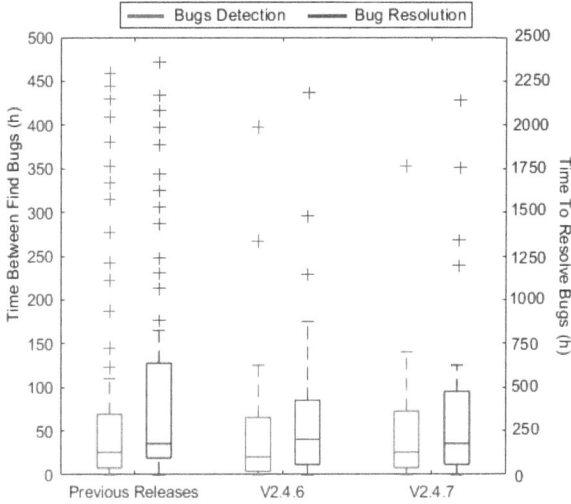

Fig. 5. Median of Bug Detection and Resolution in Hyperledger Fabric

with previous models for residual bugs in Ethereum and Hyperledger Fabric is shown in Figs. 6 and 7. The proposed model, focusing on residual bugs and their resolution, demonstrates improved performance and reliability.

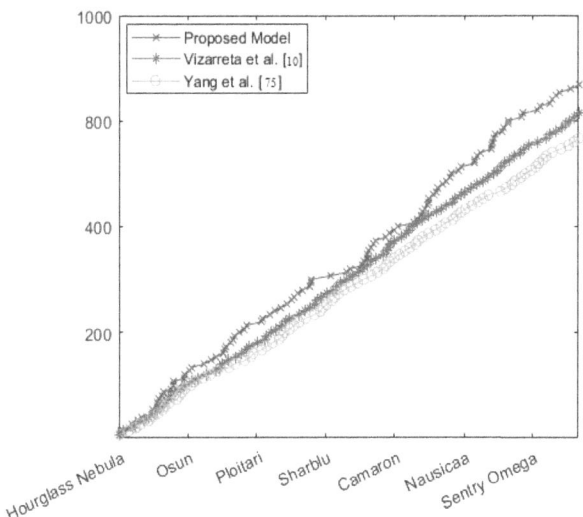

Fig. 6. Comparison of Proposed Model with Previous Model for Residual Bugs in Ethereum

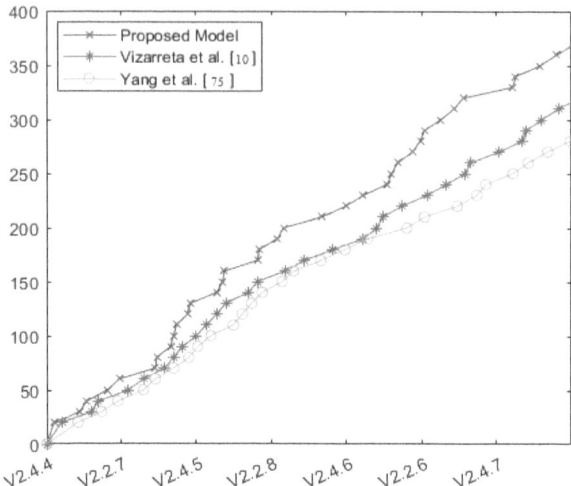

Fig. 7. Comparison of Proposed Model with Previous Model for Residual Bugs in Hyperledger Fabric

7 Discussion

The experimental results highlight several key findings: Effectiveness of SRGMs: Different SRGMs showed varying levels of effectiveness in detecting and predicting bugs across different releases. No single SRGM was universally superior, indicating the need for tailored models for specific contexts.

Residual Bug Focus: The focus on residual bugs in the proposed model improved the accuracy and relevance of bug detection and resolution predictions. This focus is critical for enhancing the maturity and reliability of blockchain-based implementations.

Model Performance: The proposed model outperformed previous Ethereum and Hyperledger Fabric models, as evidenced by the GOF metrics and comparative analysis.

Continuous Improvement: The analysis of bug detection and resolution over multiple releases shows the importance of continuous monitoring and improvement in software reliability.

These findings underscore the value of applying and refining SRGMs for blockchain-based implementations, contributing to more reliable and mature software systems.

8 Conclusion

This research uses Software Reliability Growth Models (SRGMs), particularly S-shaped SRGMs, to assess software reliability in blockchain-based applications. Two open-source datasets, Ethereum and Hyperledger Fabric, are utilized. The selected SRGMs utilize bug reports to estimate future bugs, with Least Square Estimation (LSE) employed to evaluate the goodness of fit (GOF). Three parameters are proposed for GOF assessment: Mean Square Error (MSE) for selecting appropriate SRGMs, Theil's Statistics (TS) for comparing SRGM fitness across releases, and Coefficient of Determination for variance analysis. The proposed model performs superior to existing goodness models, offering accurate cost estimation for residual bugs. We believe that the proposed model could help practitioners assess release maturity by tracking residual bugs across releases and is effective in monitoring bug introduction and removal.

References

1. Liu, Z., Kang, R.: Imperfect debugging software belief reliability growth model based on uncertain differential equation. IEEE Trans. Reliab. **71**, 735–746 (2022). https://doi.org/10.1109/TR.2022.3158336
2. Wayne, M., Modarres, M.: A Bayesian model for complex system reliability growth under arbitrary corrective actions. IEEE Trans. Reliab. **64**, 206–220 (2015). https://doi.org/10.1109/TR.2014.2337072
3. Singh, V.B., Sharma, M., Pham, H.: Entropy based software reliability analysis of multi-version open source software. IEEE Trans. Software Eng. **44**, 1207–1223 (2018). https://doi.org/10.1109/TSE.2017.2766070
4. Moranda, P.B.: An error detection model for application during software development. IEEE Trans. Reliabil. **30**, 309–312 (1981). https://doi.org/10.1109/TR.1981.5221096
5. Tandon, A., Kaur, P., Mäntymäki, M., Dhir, A.: Permissioned vs. permissionless blockchain: how and why there is only one right choice. J. Software US **16**, 95–106 (2021). https://doi.org/10.1016/j.techfore.2021.120649
6. "What is Ethereum? — ethereum.org" ethereum.org 2022. https://ethereum.org/en/what-is-ethereum. Accessed 4 Oct 2022
7. Gaba, P., Raw, R.S., Mohammed, M.A., Nedoma, J., Martinek, R.: Impact of block data components on the performance of blockchain-based VANet implemented on hyperledger fabric. IEEE Access **10**, 71003–71018 (2022). https://doi.org/10.1109/ACCESS.2022.3188296
8. Hui, Z., Liu, X.: Research on software reliability growth model based on gaussian new distribution. Procedia Comput. Sci. **166**, 73–77 (2020). https://doi.org/10.1016/j.procs.2020.02.019
9. Li, Q., Pham, H.: A generalized software reliability growth model with consideration of the uncertainty of operating environments. IEEE Access **7**, 84253–84267 (2019). https://doi.org/10.1109/ACCESS.2019.2924084
10. Liu, Z., Kang, R.: Imperfect debugging software belief reliability growth model based on uncertain differential equation. IEEE Trans. Reliab. **71**(2), 735–746 (2022). https://doi.org/10.1109/TR.2022.3158336
11. Liu, Z., Yang, S., Yang, M., Kang, R.: Software belief reliability growth model based on uncertain differential equation. IEEE Trans. Reliab. **71**(2), 775–787 (2022). https://doi.org/10.1109/TR.2022.3154770
12. Hou, Y.-F., Huang, C.-Y., Fang, C.-C.: Using the methods of statistical data analysis to improve the trustworthiness of software reliability modeling. IEEE Access **10**, 25358–25375 (2022). https://doi.org/10.1109/ACCESS.2022.3154103

13. Wu, C.Y., Huang, C.Y.: A study of incorporation of deep learning into software reliability modeling and assessment. IEEE Trans. Reliab. **70**(4), 1621–1640 (2021). https://doi.org/10.1109/TR.2021.3105531
14. Li, N., et al.: Standardization workflow technology of software testing processes and its application to SRGM on RSA timing attack tasks. IEEE Access **10**, 82540–82559 (2022). https://doi.org/10.1109/ACCESS.2022.3196934
15. Garg, R., Raheja, S., Garg, R.K.: Decision support system for optimal selection of software reliability growth models using a hybrid approach. IEEE Trans. Reliab. **71**(1), 149–161 (2022). https://doi.org/10.1109/TR.2021.3104232
16. Xiong, H., et al.: On the design of blockchain-based ECDSA with fault-tolerant batch verification protocol for blockchain-enabled IoMT. IEEE J. Biomed. Health Inform. **26**(5), 1977–1986 (2022). https://doi.org/10.1109/JBHI.2021.3112693
17. Chen, R., et al.: BIdM: a blockchain-enabled cross-domain identity management system. J. Commun. Inf. Networks **6**(1), 44–58 (2021). https://doi.org/10.23919/JCIN.2021.9387704
18. Dai, W., Dai, C., Choo, K.K.R., Cui, C., Zou, D., Jin, H.: SDTE: a secure blockchain-based data trading ecosystem. IEEE Trans. Inf. Forensics Secur. **15**, 725–737 (2020). https://doi.org/10.1109/TIFS.2019.2928256
19. Hinarejos, M.F., Ferrer-Gomila, J.L., Barceló, A.J.: A secure solution for a blockchain-based consortium promotional scheme. IEEE Access **10**, 119676–119691 (2022). https://doi.org/10.1109/ACCESS.2022.3221424
20. Politou, E., Casino, F., Alepis, E., Patsakis, C.: Blockchain mutability: challenges and proposed solutions. IEEE Trans. Emerg. Top. Comput. **9**(4), 1972–1986 (2021). https://doi.org/10.1109/TETC.2019.2949510
21. Bao, Z., Wang, Q., Shi, W., Wang, L., Lei, H., Chen, B.: When blockchain meets SGX: an overview, challenges, and open issues. IEEE Access **8**, 170404–170420 (2020). https://doi.org/10.1109/ACCESS.2020.3024254
22. Thakker, U., Patel, R., Tanwar, S., Kumar, N., Song, H.: Blockchain for diamond industry: opportunities and challenges. IEEE Internet Things J. **8**(11), 8747–8773 (2021)
23. Vizarreta, P., et al.: Assessing the maturity of SDN controllers with software reliability growth models. IEEE Trans. Netw. Serv. Manage. **15**, 1090–1104 (2018). https://doi.org/10.1109/TNSM.2018.2848105
24. Nguyen, D.C., Pathirana, P.N., Ding, M., Seneviratne, A.: Integration of blockchain and cloud of things: architecture, applications and challenges. IEEE Commun. Surv. Tutor. **22**(4), 2521–2549 (2020). https://doi.org/10.1109/COMST.2020.3020092
25. Wu, Y.P., Hu, Q.P., Xie, M., Ng, S.H.: Modeling and analysis of software fault detection and correction process by considering time dependency. IEEE Trans. Rel. **56**(4), 629–642 (2007)

An Automated Ontology-Based Requirements Traceability Technique in Agile Software Development Context

Saif Ur Rehman Khan[1(✉)], Uswa Aslam[2], Atif Mashkoor[3], Irum Inayat[4], and Habib Un Nisa[1]

[1] Department of Software Engineering, Shifa Tameer-e-Millat University (STMU), Islamabad, Pakistan
{saif_rehman.ssc,habibunnisa.ssc}@stmu.edu.pk
[2] Department of Computer Science, Superior University, Lahore, Pakistan
uswa.aslam95@gmail.com
[3] Institute for Software Systems Engineering, Johannes Kepler University (JKU), Linz, Austria
atif.mashkoor@jku.at
[4] Department of Software Engineering, National University of Computer and Emerging Science (FAST-NUCES), Islamabad, Pakistan
irum.inayat@nu.edu.pk

Abstract. Agile software development (ASD) is a popular process in the software development industry due to its dynamic nature and ability to release software quickly. However, managing the requirements in the ASD context remains a challenging task. This is because frequent changes need to be accommodated in the iterations. Consequently. It is challenging to effectively handle the changes and trace the requirements among artifacts such as sub-user stories, tasks, models, code snippets, and test cases. On the other hand, semi-automated tool support and organizational issues further complicate the requirement management process. Motivated by this, we propose an automated ontology-based technique to address the above-mentioned challenges, which benefits requirements management in agile development. The proposed technique involves tracing requirements, ensuring ontology consistency using competency questions, prioritizing user stories based on business value, and estimating the effort required for each user story. We empirically assessed the performance of the proposed technique by comparing it with baseline approaches using IEEE standards. The attained results indicate that the proposed technique significantly outperforms the baseline approaches. We believe that the proposed technique provides a cost-effective solution for requirement traceability in ASD.

Keywords: Software Requirement Management · Requirement Traceability · Ontology · Agile Software Development

---

The work of Atif Mashkoor is supported by the Austrian Science Fund (FWF) grant # I 4744-N and the LIT Secure and Correct Systems Lab sponsored by the province of Upper Austria.

1 Introduction

The traditional requirement engineering relies on thorough documentation before project initiation. At the same time, Agile methodologies prioritize individuals and interactions, working software, and customer collaboration, allowing for flexible scope and requirement definitions reviewed and re-prioritized iteratively within Scrum methodology. In the case of software development, requirements engineering is regarded as an important software development phase. Requirements for software products are inherently complex because they are based on information gathered from various sources during the requirements engineering process.

Managing evolving requirements is impossible without traceability. To address this issue, the project must meet traceability properties [1]. Agile Software Development (ASD) provides the ability to adapt to changing requirements while working on a product. However, the main challenges in enabling traceability are inefficient communication and hurdles to obtaining crucial information [2].

Ontologies offer a structured approach to managing requirements by providing domain-specific knowledge and facilitating stakeholder communication. They enhance traceability by describing requirement artifacts and enabling automated reasoning, quicker navigation, and relationship updating. However, the literature lacks sufficient focus on requirement traceability in agile development, particularly in ontology. Moreover, existing methodologies cannot estimate effort, prioritize user stories, and ensure consistency in requirements [3,4,6,7,9].

Inspired by this, the current work proposes an automated ontology-based technique that effectively traces the requirements in the ASD context. The main Research Contributions (RCs) of this work are as follows:

- RC1: Presents an ontology-based requirement traceability approach for agile methods.
- RC2: Develops and evaluates the competency questions using description logic queries to ensure ontology consistency.
- RC3: Prioritizes the user stories based on their business value and urgency.
- RC4: Estimates the effort required to complete each user story in the product backlog.
- RC5: Empirically evaluates the performance of the proposed approach using a clean methodology and empirically comparing it with the baseline approaches using IEEE standards.

The remaining sections of this paper are organized as follows: Sect. 2 provides the related work. Section 3 presents the proposed technique, and Sect. 4 presents the ontology evaluation methodology. Finally, Sect. 5 concludes this work and outlines potential future research directions.

2 Related Work

This section provides current state-of-the-art ontologies to support requirement traceability in the Agile Software Development (ASD) context. An ontology-based approach called Web Ontology Language (OWL) has been proposed in [3]. The ontology captures the unique characteristics of managing requirements in Agile projects, facilitating tracing relationships between different artifacts. The authors argued that their proposed approach effectively managed requirement growth; however, it does not consider the evaluation measures for user perspectives.

In contrast, Murtazina and Avdeenko [4] proposed an approach covering various aspects of managing software requirements in Agile settings. Their ontology reflects user narrative structures and incorporates matrices to assess user story quality, priority, and risks. However, their proposed approach does not employ a validation mechanism.

Another technique introduced in [5] integrates ontology activities into Scrum Agile methodology, resulting in the Agile Methodology for Ontology Development (AMOD) framework. AMOD facilitates ontology development across pre-game, development, and post-game stages. The authors demonstrated that Agile approaches streamline ontology development tasks. However, their proposed approach does not adopt new applications for additional validation cases.

Moreover, Cleland-Huang [21] introduced the concept of Just-in-time Traceability (JITT) as a cost-effective approach for traceability in agile projects, which supports automatic trace generation as needed to reduce workload and increase efficiency. Furtado and Zisman [22] proposed Trace++, a traceability technique that extends traditional traceability relationships to support transitioning from traditional to agile software development. It addressed problems in agile projects such as lack of metrics for rework measurement, interruption of projects, requirement changes, lack of documentation, and management control. Junior et al. [12] proposed the Scrum Reference Ontology (SRO) to address data dispersion issues in agile development applications, enabling seamless integration and data sharing. By utilizing SRO to integrate Azure DevOps and Clockify, the software artifacts facilitated improved estimates, team allocation, productivity management, project performance, and issue resolution in the Scrum process.

Table 1 provides a brief description of the current state-of-the-art regarding the employed methodology, considered dataset, evaluation measures, and limitations.

Table 1. Literature Review Matrix

Ref	Description	Methodology/ Techniques	Dataset Used	Evaluation measures	Limitations
[3]	Proposed an ontology approach using OWL.	Cross-sectional Questionnaire Quantitative	Data of 71 RE working in an Agile environment.	Nil	Deficiency in evaluation measures to consider the users' perspectives.
[4]	Developed OWL ontology to amass information about the criteria for evaluating the quality of user stories, requirements artifacts, and other kinds of requirements.	Experimental Study Qualitative	User stories data	Matrices are used to access risks in practice.	No validation was performed. The data used in this paper is unverified.
[7]	Proposed the ontology called OntoAgile, which aims to establish a unified terminology for formally sharing knowledge on the agile approach.	Mixed-Method Approach	Competency questions data	Interviews with agile practitioners and software engineering specialists were performed.	Needs to improve and expand the capacity of the assessment process through automation. Need to evaluate this ontology using case studies
[8]	Proposed requirements traceability ontology (RTOnto) to support requirements management activities in large software development.	Mixed-Method Approach, Quantitative Approach	Experiments case study data of the university library		This study evaluated software development in a university library using a case study. It lacks checking the semantic consistency of artifacts, requirement traceability, and reasoning.
[6]	Proposed a new approach for change request classification based on ontology.	Experimental Study Quantitative	Collected users' reviews from the PROMISE repository	Reasoning features	This study does not provide a comparison with existing classification methods or tools to showcase the superiority of the proposed approach
[10]	Proposed the Requirements Traceability Ontology (RTOnto) to support requirements management activities in software development, focusing on flexible trace granularity and classification of artifacts into eight sub-classes	Case Study	User stories data	Evaluated using a case study of software development in a university library	This study does not explicitly focus solely on agile methodologies.
[11]	the proposed collaborative framework based on Agile and Scrum models provided system features for supporting ontology development	Experimental Study	Use cases data related to Budha images	Collaborative use case scenario	The study does not comprehensively address miscommunication in cross-disciplinary concept discovery.

3 Proposed Ontology-Based Requirement Traceability Technique

This section presents the proposed ontology-based technique for effective requirements traceability in the context of Agile Software Development (ASD). Figure 1 illustrates an abstract view of the proposed technique. The proposed technique consists of five phases:

- Phase 1: Performs requirements traceability
- Phase 2: Competency questions
- Phase 3: Ranks the user stories
- Phase 4: Estimates the required efforts of user stories
- Phase 5: Validates the proposed technique

As previously discussed, we are dealing with agile methods in which requirements are written as user stories. The related tasks will be considered sub-classes of the "requirement." This category contains sub-classes like "acceptance criteria." In addition, acceptance criteria and "definition of done" are also components of the class "requirement artifact."

The requirements and artifacts are made up of object and data properties. Table 2 depicts a top-level class taxonomy, a portion of object characteristics, and an example of axioms. Classes (requirements, stakeholders, artifacts, project backlog, risks, sprints, user stories, Tasks, Priority Factors, and acceptance criteria) sub-classes and their descriptions.

Table 2. Description of upper-level classes

Classes	Description
Requirements	Include subclasses: Functional Requirement, Non-Functional Requirements
Stakeholder	Contains subclasses corresponding to the concept "stakeholder"
Artifacts	Include subclasses describing the artifacts: Scrum Artifacts, Requirement Artifact, User-story-Artifacts
Project Backlog	Includes subclasses describing the project's backlogs: maintenance backlog, product backlog, sprint backlog, task backlog
Risks	Includes subclasses describing the classification of risks
Sprint	Contains subclasses describing the sprint planning, Release, Sprint Backlog, Sprint Retrospective
User Stories	Includes subclasses i.e., User Story dependency, User Story Priority,
Task	Includes instances which are tasks a development team works on
Priority Factor	Contains subclasses related to "Priority Factors"
Acceptance Criteria	Contains subclasses corresponding to the concept "Acceptance Criteria"

3.1 Traceability of Requirements Knowledge in an Ontology

An ontology is a database containing data on requirements, tasks, acceptance criteria, scrum artifacts, requirements sources (including stakeholders), risks,

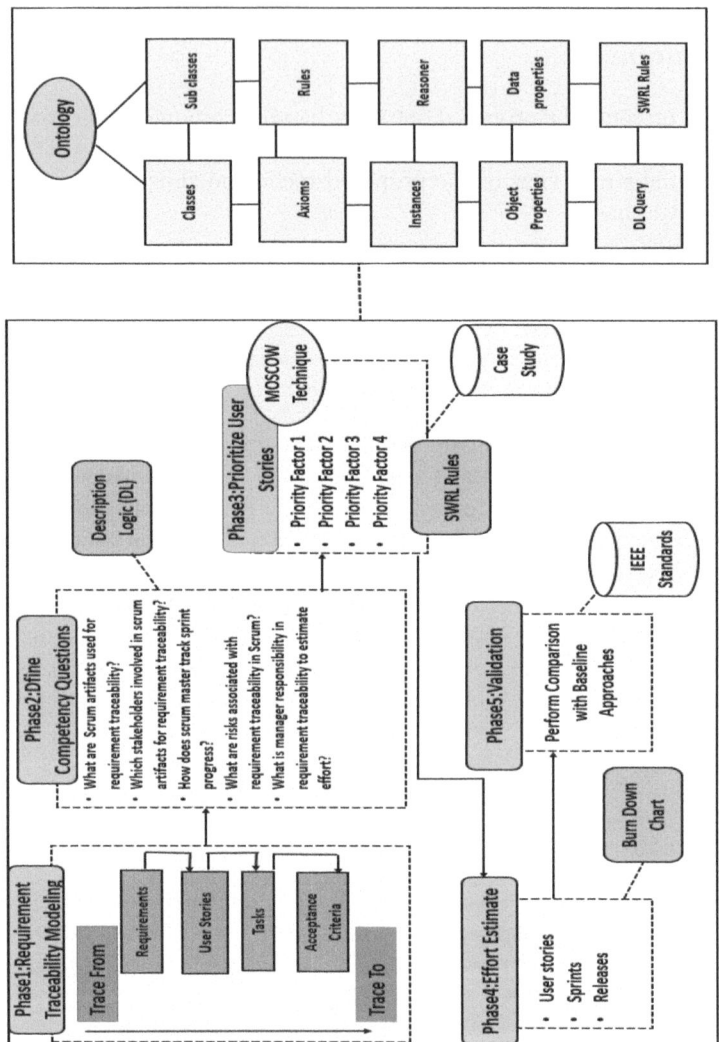

Fig. 1. High-Level View of Proposed Ontology-Based Requirement Traceability Technique

priority factors, and team members who add requirements, including knowledge, to the ontology. When modifying the software requirements, the relationships between the instances can be used to identify requirements and artifacts directly or indirectly affected.

The ontology includes two types of relations implemented through object properties: (i) "traceFrom" and (ii) "traceTo," as shown in Fig. 2. The first relation includes sub-relations that enable bottom-up tracing and the second top-down tracing. Subrelations of the "relation user stories" are:

(1) As an (intended user)
(2) I want to (intended action)
(3) so that (outcome of the action)

The relation_task enables finding tasks that check a functional part of a software product. Subrelation of the relation_task are:

(1) In progress (task in progress)
(2) To _do (task can work on)
(3) To verify (task pending for verification or testing).

The relation acceptance criteria provide detailed functionality details that assist the team in determining whether the story is complete and functions as planned. Subrelation of the relation acceptance criteria are:

(1) Given (How things begin)
(2) Then (outcome of taking action)"
(3) Then (Action taken).

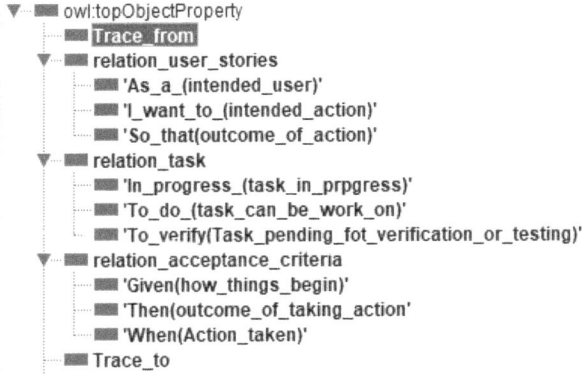

Fig. 2. Traceability Relationship "Trace-From" to "Trace-to"

3.2 Usage View of Ontology

The ontology also includes axioms that verify pattern matching between user stories and behavior situations. The Usage view displays axioms that reference (or mention) the currently selected entity. It displays axioms containing the selected entity's identity. All expressions are hyperlinked for easy browsing and sorted by a topic entity (where possible). The Usage view can assist in determining how frequently a particular item is used in an ontology or where a class has been used as a restriction filler. Figure 3 described the usage view of trace from object property. The usage view of the "Trace from" object property displays the Domain, Range, and Sub properties.

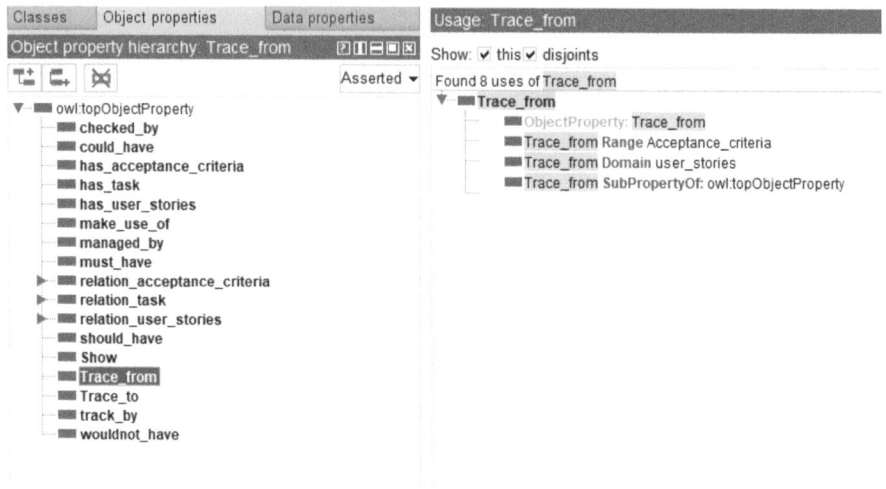

Fig. 3. Usage View of "Trace From" Property

3.3 Using a Reasoner to Test the Ontology

A reasoner is essential for maintaining consistency and ensuring high-quality data storage in an ontology. It evaluates class relationships, such as determining if one class is a subclass of another, which helps to detect errors and inconsistencies [13,14]. For instance, applying a reasoner to the ontology in this study revealed that it's necessary for requirements. It ensures each class contains relevant instances and detects inconsistencies by checking consistency constraints and instance types. Moreover, a reasoner automatically computes the class hierarchy, eliminating the need for manual work.

3.4 Description Logic (DL) Query in the Requirement Traceability

Using a DL Query tab enables the user to rapidly examine the definition of a class to identify the appropriate subclass. The DL Query tab includes an easy-to-use feature for searching a categorized ontology and the ability to process the ontology and look for relationships. Various links include direct superclasses, equivalent classes, and subclasses. DL Query also supports using a reasoner to prevent ontology conflicts [15]. The methodology employed in this study consists of a list of questions from which the user chooses a response based on the ontology's response. Figure 4 shows the list of competency questions about the various requirement artifacts, scrum artifacts, risks, manager responsibility, scrum master, and requirements. A description logic query is generated against each competency question, and the query results of each question are also shown.

Automated Ontology-Based Requirements Traceability Technique 37

Competency Questions	DL qurey	Query results
What are Scrum artifacts used for requirement traceability?	Scrum artifacts "**make use of** "	User story Task Sprint backlog Scrum board Definition of done Velocity chart Burn down chart Product increment
Which stakeholders involved in scrum artifacts for requirement traceability?	Scrum artifacts "**checked by**"	Manager developer Tester product owner Scrum master Agile team
How does scrum master track sprint progress?	"**Track by**" scrum master	Daily scrum meeting scrum retrospective sprint planning defect density sprint burn down
What are risks associated with requirement traceability in Scrum?	"**show**" list of risks	Budget people team sprint(duration ,deliverable) Product(user stories, epics tasks) Knowledge and capabilities
What is manager responsibility in requirement traceability to estimate effort?	"**managed by**" by manager	Test coverage Test assignment Test execution

Fig. 4. Competency Questions related to Description Logic Query

3.5 Semantic Web Rule Language (SWRL) Rule for Requirement Traceability

During sprint planning, backlog prioritization establishes the order for developing and implementing product backlog items (such as user stories, defects, and spikes). The Scrum team utilizes this process to select the items.

SWRL, a Semantic Web rule language, is represented by OWL concepts. The ontology contains classes and their characteristics. Figure 5 shows rules for generating risk values. Notice that the letters in Fig. 5, followed by a question mark, indicate values for specific classes. SWRL extends OWL, providing a semantic layer to express statements that OWL cannot handle [16].

Priority Factors	SWRL Rules
PF4	user_stories (? u) ^have_limit (?u ,?f l) ^ swrlb: lessThanOrEqual (? l, 25) ^must_have(?u, ?w) -> PF4(?u)
PF3	user_stories (? u) ^have_limit (?u, ?f) ^ swrlb:GreaterThanOrEqual(?f, 15) ^should_have(?u, ?g) -> PF3(? u)
pF2	user_stories (? u) ^ have_limit(?u, ?f) ^ swrlb:GreaterThan(?f, 5) ^could_have(?u, ?g) -> PF2(?u)
PF1	user_stories (? u) ^have_limit(?u, ?f) ^ swrlb:lessThan(?f, 6) ^ wouldnot_have(?u, ?g) ->PF1(?u)

Fig. 5. SWRL Rules and Priority Factors

3.6 Prioritization of User Stories in Product Backlog

In agile development, various metrics prioritize user stories, with Story Points and Effort being among the most common. However, our approach, which uses Urgency and Business Value, is well-founded and complements traditional metrics. Urgency captures user stories' time sensitivity and dependency aspects, critical for aligning development efforts with deadlines and interdependent tasks. Business Value emphasizes the potential impact on the business, ensuring that the development team focuses on features that deliver the most significant benefits to stakeholders.

To prioritize user stories in the product backlog, we calculate the priority using the following equation:

$$\text{Priority} = \text{Urgency} \times \text{Business Value} \qquad (1)$$

Urgency The urgency of a user story is determined by the product owner's deadline, classified as follows:

- **5 - Extremely Urgent:** Time-consuming and heavily dependent on other factors.
- **4 - Very Urgent:** Time-constrained due to customer or contextual requirements, highly dependent on other items.
- **3 - Moderately Urgent:** Moderate time commitment, somewhat dependent on other tasks.
- **2 - Slightly Urgent:** Minimal time requirements, little dependence on other items.
- **1 - Not Urgent:** No time constraint or dependency.

Business Value. Business value refers to the potential income gained or lost from a user story, classified as follows:

- **5 - Very High:** Critical to most customers, greatly influences brand or business effectiveness.
- **4 - High:** Key factors for many customers significantly impact reputation or brand.
- **3 - Moderate:** Significant to a smaller customer base, moderate impact on reputation or brand.
- **2 - Low:** Important only to new customers, minimal influence on reputation or brand.
- **1 - Very Low:** Minimal importance to customers, negligible impact on reputation or brand.

Priority factors (PF1, PF2, PF3, PF4) are assigned based on urgency and business value, with PF4 being the highest priority and PF1 the lowest. These factors are determined using SWRL (Semantic Web Rule Language) to ensure effective prioritization of user stories in the product backlog.

3.7 Effort Estimation of User Stories

A burndown chart is a visual representation tracking the progress of work completed and remaining within a sprint or epic. It helps Scrum teams estimate effort, monitor progress, and detect scope creep. During sprint planning, teams use the chart to estimate achievable work. The chart, often created with tools like MS Excel, shows remaining effort and task completion trends. The Scrum Master and Product Owner use it to track progress and intervene if needed, ensuring the team stays on track to meet their sprint commitments.

4 Ontology Evaluation Methodology

This section introduces the evaluation methodology to support this study. The proposed technique is evaluated using onClean methodology [20] and IEEE standard [18].

4.1 Ontology Evaluation Using OntoClean

Ontologies' effectiveness is evaluated based on correctness, task orientation, completeness, conciseness, expandability, reusability, and clarity.

Correctness: Ensures all elements adhere to guidelines and every inference is true.

Task Orientation: Determines if the ontology meets its development conditions.

Completeness and Conciseness: Contains all required information without unnecessary elements.

Expandability and Reusability: Assesses ease of adding functionalities and reusing components. Clarity: Ensures the ontology is understandable and easy to analyze.

The OntoClean methodology evaluates requirement traceability in agile development. It checks if class instances adhere to criteria, their uniqueness, and dependencies on external concepts. In Fig. 6, +R indicates that instances of the class Requirement will always belong to this concept, +I shows unique identification criteria, -D means no external dependencies, and +D indicates a dependency on an external concept. +U signifies that instances of the class are available as "Whole."

4.2 Ontology Evaluation Using IEEE Standards

Evaluation based on IEEE standards [18] involves comparing the requirement traceability methods against predefined processes such as project management, ontology development, and integral processes. The analysis rates each approach's compliance with IEEE processes as Covered (C), Partially Covered (P), or Uncovered (U). The current ratings were

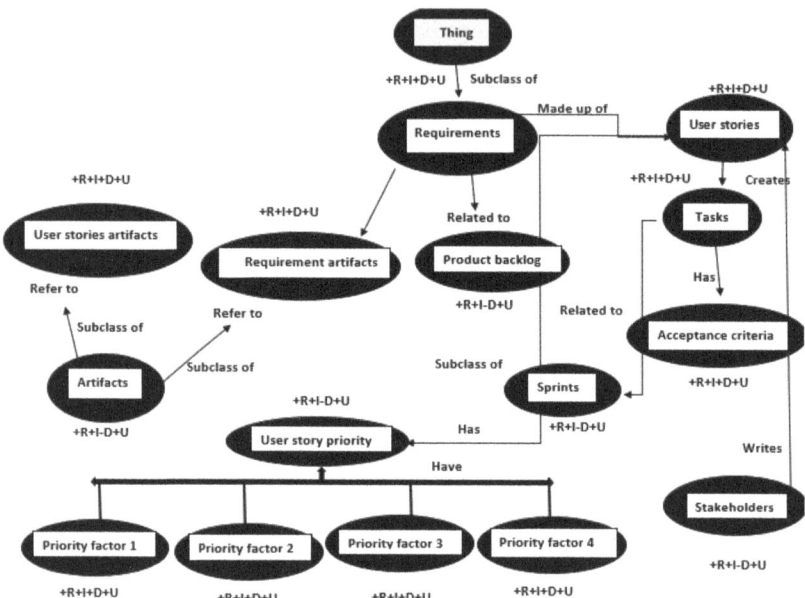

Fig. 6. Requirement Traceability in Ontology Evaluation through Onto Clean

calculated using the following criteria: the number of processes $X_{r,d}$ and approach a, (a = 1, .. m), For specific rating r(r = 1 .. n) [17].

$$P_r = \frac{X_{r,d}}{\sum_{r=1}^{n} X_{r,d}} \times 100. \qquad (2)$$

Based on the evaluation, existing approaches lack comprehensive coverage of project management processes, including risk management and scheduling. Most approaches focus more on development activities and overlook essential project management aspects. The proposed technique improves project management, integrative processes, and essential operations in ontology construction. It receives a 70% satisfaction and 22% partial satisfaction rating for IEEE standard. Overall, existing approaches show varying levels of compliance with IEEE standards. The proposed approach substantially improves ontology construction and requirement traceability in agile development environments.

4.3 Significance of Proposed Approach

The proposed ontology-based technique for requirements traceability offers several significant advantages in the context of Agile Software Development (ASD). This approach ensures comprehensive traceability and seamless integration of requirements with project artifacts by structuring requirements as user stories and defining their relationships through a well-designed ontology. For example, in an Agile project developing a new e-commerce platform, user stories such as

"As a customer, I want to filter products by category so that I can find items easily" are linked to acceptance criteria and tasks. The ontology's traceability feature allows team members to trace dependencies and impacts, such as how changes to the filtering functionality might affect the product catalog or search results. The ranking phase prioritizes user stories based on urgency and business value, which ensures that critical features like secure payment integration are developed first. Effort estimation is visualized through burndown charts, which give the team clear progress indicators and help the Scrum Master make timely interventions. The ontology-based approach enhances clarity and consistency and supports dynamic decision-making and prioritization, ultimately leading to a more efficient and effective development process. The proposed technique mitigates the risk of missing requirements and facilitates adaptive planning, which is important for the iterative nature of Agile methodologies.

5 Conclusion and Future Work

This work introduced an ontology-based technique to enhance requirement traceability in agile software development. As a result, it is beneficial in addressing gaps in effort estimation, user story prioritization, and consistency checking. The proposed technique facilitates requirement traceability by gathering information about requirements, employing competency questions, prioritizing user stories, estimating effort using burn-down charts, and comparing existing ontology engineering approaches against IEEE standards. The results indicate that the proposed technique aligns more closely with IEEE standards than other approaches, especially when evaluated using the OntoClean methodology.

In the future, we plan to integrate different ontologies, employ machine learning for conflict identification, extend the application to various requirement engineering domains, and incorporate formal specifications into natural language requirements. Moreover, automating the assessment process through algorithm development could be a potential research dimension for future researchers.

References

1. Dai, P., Yang, L., Wang, Y., Jin, D., Gong, Y.: Constructing traceability links between software requirements and source code based on neural networks. Mathematics **11**(2), 315 (2023)
2. Salem, A.M.: Model for enhancing requirements traceability and analysis. Int. J. Adv. Comput. Sci. Appl. **1**(5), 14–21 (2010)
3. Murtazina, M.S., Avdeenko, T.V.: An ontology-based approach to support for requirements traceability in agile development. Procedia Comput. Sci. **150**, 628–635 (2019)
4. Murtazina, M.S., Avdeenko, T.V.: Ontology-based approach to the requirements engineering in agile environment. In: 2018 XIV International Scientific-Technical Conference on Actual Problems of Electronics Instrument Engineering (APEIE), pp. 496–501. IEEE (2018)

5. Abdelghany, A.S., Darwish, N.R., Hefni, H.A.: An agile methodology for ontology development. Int. J. Intell. Eng. Syst. **12**(2), 170–181 (2019)
6. Sakhrawi, Z., Sellami, A., Bouassida, N.: Requirements change requests classification: an ontology-based approach. In: Abraham, A., Siarry, P., Ma, K., Kaklauskas, A. (eds.) ISDA 2019. AISC, vol. 1181, pp. 487–496. Springer, Cham (2021). https://doi.org/10.1007/978-3-030-49342-4_47
7. Murtazina, M.S., Avdeenko, T.V.: The ontology-driven approach to support the requirements engineering process in Scrum framework. In: CEUR Workshop Proceedings, vol. 2212, pp. 287–295 (2018)
8. Bjørner, D., Broy, M., Pottosin, I.V. (eds.): PSI 1996. LNCS, vol. 1181. Springer, Heidelberg (1996). https://doi.org/10.1007/3-540-62064-8
9. Ortega-Ordonez, W.A., Pardo-Calvache, C.J., Pino-Correa, F.J.: OntoAgile: an ontology for agile software development processes. Dyna **86**(209), 79–90 (2019)
10. Wibowo, A., Davis, J.: Requirements traceability ontology to support requirements management. In: Proceedings of the Australasian Computer Science Week Multi-conference, pp. 1–9 (2020)
11. Takhom, A., Usanavasin, S., Supnithi, T., Boonkwan, P.: A collaborative framework supporting ontology development based on agile and scrum model. IEICE Trans. Inf. Syst. **103**(12), 2568–2577 (2020)
12. Júnior, P.S.S., Barcellos, M.P., de Almeida Falbo, R., Almeida, J.P.A.: From a scrum reference ontology to the integration of applications for data-driven software development. Inf. Softw. Technol. **136**, 106570 (2021)
13. Horridge, M., Jupp, S., Moulton, G., Rector, A., Stevens, R., Wroe, C.: A practical guide to building owl ontologies using protege 4 and co-ode tools edition1. 2. The university of Manchester, 107 (2009)
14. Fahad, M., Qadir, M.A., Shah, S.A.H.: Evaluation of ontologies and DL reasoners. In: Shi, Z., Mercier-Laurent, E., Leake, D. (eds.) IIP 2008. ITIFIP, vol. 288, pp. 17–27. Springer, Boston, MA (2008). https://doi.org/10.1007/978-0-387-87685-6_5
15. Khamparia, A., Pandey, B.: Performance analysis of SPARQL and DL-QUERY on electromyography ontology. Indian J. Sci. Technol. **8**(17), 2015 (2015)
16. Skillen, K.L., Chen, L., Nugent, C., Donnelly, M., Burns, W., Solheim, I.: Using SWRL and ontological reasoning for the personalization of context-aware assistive services. In: Proceedings of the 6th International Conference on PErvasive Technologies Related to Assistive Environments, pp. 1–8 (2013)
17. Zahraoui, H., Idrissi, M.A.J.: Adjusting story points calculation in scrum effort time estimation. In: 2015 10th International Conference on Intelligent Systems: Theories and Applications (SITA), pp. 1–8. IEEE (2015)
18. Ferna'ndez-Lo'pez, M., Go'mez-Pe'rez, A.: Overview and analysis of methodologies for building ontologies. Knowl. Eng. Rev. **17**(2), 129–156 (2002)
19. Murtazina, M., Avdeenko, T.: An ontology-based approach to the agile requirements engineering. In: Bjørner, N., Virbitskaite, I., Voronkov, A. (eds.) PSI 2019. LNCS, vol. 11964, pp. 205–213. Springer, Cham (2019). https://doi.org/10.1007/978-3-030-37487-7_17
20. Guarino, N., Welty, C.: Evaluating ontological decisions with OntoClean. Commun. ACM **45**(2), 61–65 (2002)

21. Cleland-Huang, J.: Traceability in agile projects. In: Cleland-Huang, J., Gotel, O., Zisman, A., et al. (eds.) Software and Systems Traceability, pp. 265–275. Springer, London (2011). https://doi.org/10.1007/978-1-4471-2239-5_12
22. Furtado, F., Zisman, A.: Trace++: a traceability approach to support transitioning to agile software engineering. In 2016 IEEE 24th International Requirements Engineering Conference (RE), pp. 66–75. (2016)IEEE

Toward a Knowledge-Based Anomaly Identification System for Detecting Anomalies in the Smart Grid

Sarita Paudel[1(✉)] and Abdelkader Magdy Shaaban[2]

[1] Department of Science and Technology, IMC University of Applied Sciences, Krems, Austria
`sarita.paudel@fh-krems.ac.at`
[2] Center for Digital Safety and Security, AIT Austrian Institute of Technology, Vienna, Austria
`abdelkader.shaaban@ait.ac.at`

Abstract. State Estimation (SE) supports Situation Awareness for Cyber-Physical Systems (CPS). Time-synchronized Phasor Measurement Unit (PMU) measurements are used for SE in a Smart Grid. Data injection attacks on PMU measurements can lead to incorrect estimation of a system's state and decrease the trustworthiness of SE. Anomaly detection and root cause analysis can help apply appropriate mitigations to the power system. Executing anomaly identification on measurements before estimating the states of a system makes the estimated states trustworthy. In this work, we propose a Knowledge-Based Anomaly Identification System (KBAIS) for detecting an anomaly and identifying its root cause. An ontology, a knowledge graph, and context are used for analysis. We also present a case study - applying the KBAIS to Ecole Polytechnique Federale de Lausanne (EPFL) electrical network. It illustrates anomaly identification in PMU measurements in the context of the EPFL PMUs network.

Keywords: Cyber-Physical System · State Estimation · Cyber Attacks · Knowledge base · Anomaly Identification

1 Introduction

A Smart Grid (SG) integrates Information and Communications Technology (ICT) and incorporates new functions into electricity grid monitoring and control systems. Operators use a system to monitor and control CPS in the power domain. The control system needs to be trustworthy and aware of the current situation. Awareness of the current situation helps avoid the negative impacts of cyber-attacks on the power system.

Wide Area Monitoring Systems (WAMSs) [1] provide real-time monitoring of the system by measuring synchro-phasor data from different locations. Global Positioning System (GPS) synchronized distributed PMUs provide accurate measurements of voltage and current phasors (amplitudes, phase angles)

and frequencies [2]. WAMSs use the PMU measurements as inputs to various monitoring and control applications in the grid.

Highly dynamic measurements gathered from the distributed PMUs used for SE reflect the dynamic performance of a power system in real-time. Data injection attacks in PMU measurements can lead to incorrect SE and invoke wrong control actions, which can have a negative impact, such as life-threatening. For instance, disrupting grid operation and causing a power blackout [3], causing severe damage to hardware, software, and machinery breakdowns in a smart grid [4]. Therefore, an operator needs to know the root cause of an anomaly to apply an appropriate mitigation. Anomaly detection in the early stage, identifying its cause, and applying an appropriate mitigation strategy can help avoid such a critical situation. Faults or attacks on a power system can cause an anomaly. A fault can exist due to a fake fault introduced by an attacker or a real fault (e.g., an old and noisy sensor) in the system. Thus, analyzing a power system's behavior helps us detect the cause of faults and attacks. Similarly, a transient in a power system can also cause an anomaly. Such a transient can be detected by analyzing the system's dynamic behavior with respect to context.

Furthermore, this work proposes an ontology-based model for identifying anomalies and ensuring trustworthy SE. It aims to estimate trustworthy states by executing anomaly identification before sending measurements to state estimators. The anomaly identification system detects anomalous behavior of an observation using *anomaly detection algorithms* and rationalizes the anomaly using *context*. We propose the use of a *Knowledge Based Anomaly Identification System*, which employs an *ontology* and a *knowledge graph* to detect anomalous behavior. Knowledge about the system and decisions made by anomaly detection algorithms are used for the analysis. An appropriate mitigation strategy is applied if an observation is detected as an anomaly before sending it for estimating states. Additionally, we will present an application of the model in a power domain using a case study.

This paper is structured as follows. First, we present state-of-the-art in Sect. 2. Then we present our approach in Sect. 3. Application of the proposed knowledge-based anomaly identification system is presented in Sect. 4. The paper is concluded in Sect. 5.

2 State of the Art

This section discusses an existing body of research on anomaly identification in Smart Grids. We discuss knowledge-based anomaly detection approaches and the technology used to represent them. Real-time or near real-time data is collected from distributed SG networks to improve information communication and operational technology security. NIST [5] provides an example of SG architecture, its components, communication, information collection and analysis, and situation awareness in an SG.

Protecting communication networks against new and unexpected attacks is challenging, as new vulnerabilities emerge daily and attacks become more sophisticated [6,7]. A secure architecture can support defending against attacks by providing information about the system. Gokarn et al. [8] fed real-time estimated states using the Kalman Filter to the anomaly detection method. The method detects whether the real-time data is subject to a fault or an attack. Normal and abnormal behaviors are classified using the k-NN algorithm. If the real-time data is classified as abnormal, authors check power flow equations to detect whether it is a fault or an attack. It is identified as a fault if the real-time data satisfy the equations; otherwise, it might be an attack.

In our previous work [9], we investigated whether a combination of methods can improve the anomaly detection performance. At this end in [9], we investigated the suitability of weighted voting method for combining the AD methods, and analyzed AD performance of the weighted voting method for combination. In contrast to existing work, we used weighted voting scheme in [10] which is previously used for combining machine learning methods to combine the statistical methods in SGs.

Alcaraz et al. [11] investigate SG general architecture, control technologies, communication infrastructures, requirements that need to be fulfilled by existing anomaly detection approaches, functional features of the approaches, and their applicability in the SG context. After the investigation, the authors recommend statistical and knowledge-based approaches for control centers and corporate networks (e.g., wide area network, local network) in a SG. In this work, we consider the future work recommended in [11] and propose combining statistical analysis to knowledge-based approach. The knowledge-based approach consists of domain-specific knowledge (e.g., attacks, vulnerabilities) that helps to extract detailed information about an event [12]. In addition, dynamic features of knowledge-based approaches (e.g., expert systems) make these approaches applicable in complex and dynamic contexts. For example, an expert system [13] contains simple rules, and based on these rules, different models are capable of reasoning like a human expert about the provided knowledge of threats and vulnerabilities, security policies, etc. We propose developing an EPFL electrical network-specific system by defining rules, known issues, vulnerabilities, known attacks, etc.

Knowledge can be represented as categories, subclasses of objects whose relations can be organized in a hierarchical structure [14]. It can be represented in taxonomy (even in taxonomic hierarchy), knowledge graph, ontology, etc. Many languages, such as first-order logic, propositional logic, semantic graphs, etc., can be used to discuss content and organization of knowledge. In [15], Foley et al. propose using an ontology-based scheme to manage trust in a semantic threat graph knowledge-base model. Here, we propose using ontology and a knowledge graph to identify the malicious behavior of PMU measurements and its cause to determine the appropriate mitigation strategy for addressing any such malicious activities.

3 Approach

In this section, we aim to introduce our proposed anomaly identification system, delving into its main structure and clarifying the pivotal activities for detecting intrusions.

3.1 The Structure of the Proposed Anomaly Identification System

An illustration of the structure of the proposed anomaly identification system is depicted in Fig. 1. This figure gives an overview of the high-level structure of the proposed approach. This approach has two major components:

1. data-based detection
2. knowledge-based detection.

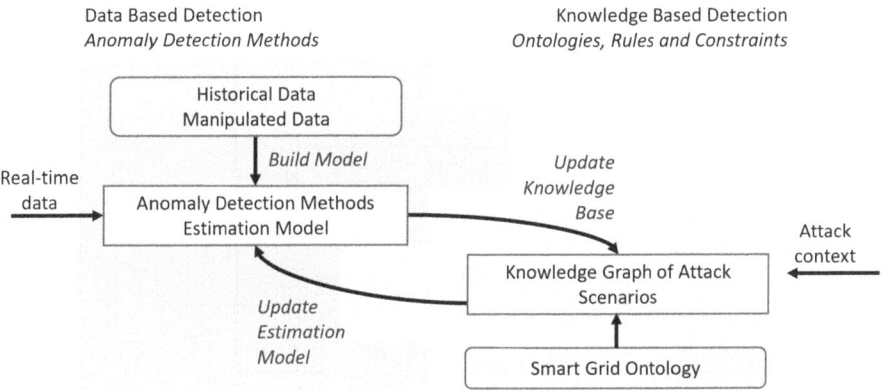

Fig. 1. The proposed anomaly identification system

The intrusion identification system has two essential components, each playing a pivotal role in the detection process:

Data-based detection uses historical data, real-time data (e.g., sensor data from smart meters and phasor measurements, network data, etc.), and statistical methods for anomaly detection. We considered different anomaly detection methods that provide insights into the decisions (i.e., their results are explainable). This is useful in intrusion detection because it allows the derivation of rules for intrusion detection and makes it robust against manipulation. Our previous work [9] demonstrates the first component (data-based anomaly detection).

Knowledge-based detection uses attack context, system context, knowledge graph, and an ontology. Due to the more deterministic nature of many machine-to-machine communications, an inclusion of knowledge about topology, devices, and protocols makes sense. Also, this can help to assess how attackers may be able to evade detection. In this work, we focus on knowledge-based detection.

3.2 The Major Activities for Detecting Intrusions

Observed measurements, context, and a normal behavior model are inputs to the anomaly identification system. Figure 2 visualizes major activities of the anomaly identification system. Information extracted from each block is presented on the side of each arrow (see Fig. 2). We name the components in the bottom left corner of each block to make it easy to recall in the next sections. The functionality of each block in this figure is briefly presented in the following paragraphs.

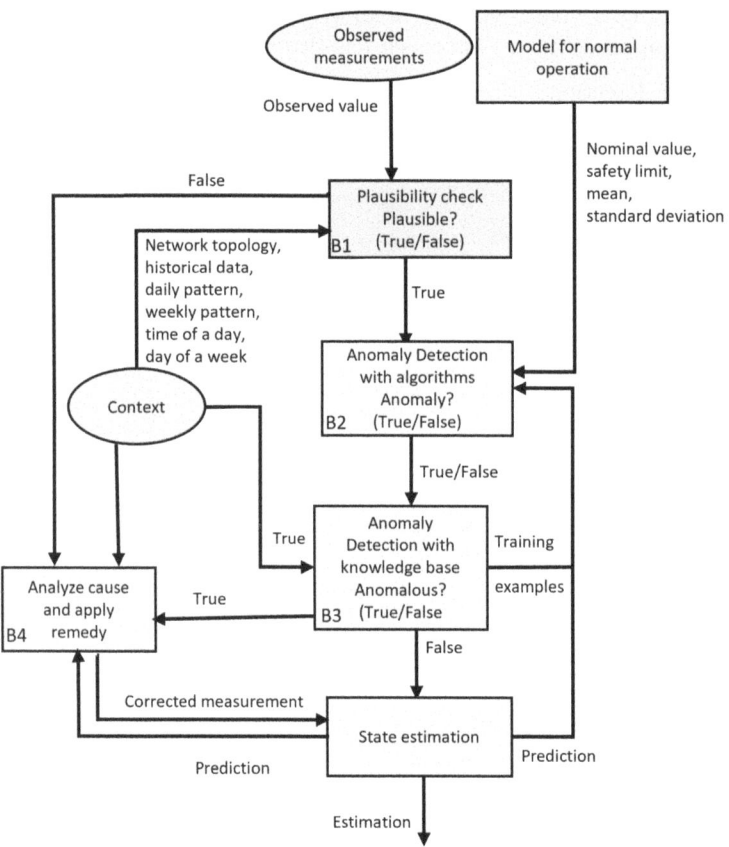

Fig. 2. An overview of major activities for detecting intrusions

Context. Available information about a system is used as input to the anomaly identification system. Network topology, historical data, time of day, day of the week, weather conditions, known system issues, etc., can be used as context. More information (e.g., data pattern) can be extracted from historical data.

Model for Normal Behavior. A normal behavior model of a system represents the behavior of the system in normal operation. Here, we develop a normal behavior model for a power system. First, we conduct an analysis of statistical features in normal operation using historical data and define their values. In the second step, we define the nominal value and general rules (e.g., safety limit) of a system in normal operation. The safety limit of a measurement according to standards is defined. For instance, in a power system, according to European Standard EN 50260 [16], voltage fluctuation in normal operation is 0.9 p.u. to 1.1 p.u. Thus, the model defines safety limit, nominal value, mean, median, and standard deviation in normal operation.

Plausibility Check. The plausibility check of an observation in B1 is carried out in two steps. First, we check the deviation of observed values from their nominal values. Domain-specific and specific rules are defined by relating multiple variables in equations. These relationships are named as physical laws of a system. In the second step, we check the variables' relationship to see whether the measured values satisfy the physical laws [17].

Other anomaly detection methods will be applied only if the observation is plausible. Otherwise, the root cause will be analyzed, and an appropriate mitigation strategy will be applied.

Anomaly Detection Using Algorithms. Different methods detect anomalies based on different features of measurements. For example, change point detection methods (e.g., Cumulative Sum (CUSUM) [18,19]), distribution-based methods (e.g., Kullback Leibler Divergence (KLD) [20,21]), SE-based methods (e.g., residual based [22]) analyze different features for anomaly detection. Sliding window mode increases anomaly detection performance. Therefore, we apply the algorithms by turning the sliding window mode on.

An anomaly that is able to bypass one anomaly detection method can be detected by another anomaly detection method. Therefore, we combine algorithms from different classes of methods to detect malicious behavior of an observation. The algorithms use information from the normal behavior model to detect whether an observation is an anomaly (B2).

We combine the algorithms to enhance anomaly detection performance. A final anomaly detection decision will be made based on the decisions of the algorithms. Several approaches can be used to make a final decision. Here, we briefly describe some of the example approaches:

- Calculate a majority vote from the decisions; the most voted decision wins. It requires an unequal number of votes; otherwise, it requires a decision rule for equal votes. Panda et al. [23] present a work un-weighted majority voting system for network intrusion detection.
- Extract known events, anomalies, and previous cyber-attacks detected by the corresponding algorithm if it detects the observation as an anomaly and analyzes its semantic graph. Mookiah et al. [24] propose a work using graph-based anomaly detection on smart homes power usage.
- Assign weight to the decisions of each algorithm based on their anomaly detection performance and use the weighted decision to calculate the probability of an observation being an anomaly. In weighted voting [25], weights are

assigned for each method. We reviewed the weighted voting method in [10] to combine the statistical methods. True negative rate and recall (true positive rate) are used for assigning weights for each method.

The combination of results is demonstrated in our previous work [9]. The final anomaly detection decision (true/false) from the combination of algorithms will be combined with context and knowledge base (B3) to investigate the behavior of an observation. The process of anomaly detection and root cause identification using contextual information and knowledge base is presented in the following subsection.

3.3 Anomaly Identification Using Context

We use contextual information to rationalize an anomaly. AD decisions from the algorithms and context are used to identify the anomalous behavior of an observation. We propose using an ontology and a knowledge graph to represent the knowledge base.

We propose to instantiate an ontology for a smart grid. It defines anomalies, anomaly types, their causes, and mitigation. An anomaly can be caused due to different reasons, e.g., sensor failure (non-malicious anomaly), data manipulation attack (malicious anomaly), connection and disconnection of load, generator, etc., which can cause imbalance and abnormal behavior (real anomaly) in the system. Therefore, anomalies are categorized into three categories based on their root cause:

– real anomaly,
– non-malicious anomaly, and
– malicious anomaly.

An anomaly can have a root cause, which many problems can cause, and each cause has at least one mitigation. For instance, in a power system, if a noisy sensor causes over-voltage, the bad data needs to be corrected, and the old sensor needs to be replaced by a new sensor. Therefore, the ontology is structured to include the type of anomaly, cause, and related mitigation.

3.4 State Estimation Using Discrete Kalman Filter

Kalman Filter [26] is widely used for SE in different domains to deal with noisy measurement data and detect Bad Data (BD) due to failures in the measurement system. SE in real-time is often used to monitor the grid and achieve situational awareness. It estimates a system state based on the previously estimated state and some additional variables (e.g., process noise, measurement noise, etc.). We use a Discrete Kalman Filter (DKF) for Linear State Estimation (LSE) of a power system. It has two stages: prediction and estimation. Prediction is based on the previous estimated state, whereas estimation of the current state is based on the prediction and observation. LSE of a power system using DKF is presented in our previous work [22]. Here, we aim to feed non-anomalous measurements to a state estimator to provide trustworthy information to operators in a control center.

4 Case Study: Anomaly in the EPFL Network

A smart grid infrastructure is deployed on the EPFL campus. The PMU measurements collected in the EPFL campus grid are available for research purposes which we use for our research. Daily patterns and weekly voltage patterns are extracted from the historical data. In addition, available information about the EPFL PMU network topology, such as loads, power generation, and sensors, are used as contextual information. EPFL electrical network has variable loads depending on the time of the day and weather conditions. Photovoltaic panels and heat and power generation units are installed for necessary power generation. For simplicity, we named them as generators while using them in the context. The voltage and current profiles are affected by the use of power electronics. 5 PMUs are deployed in the feeder and GPS, and the PMUs are synchronized in each substation. Altea CVS-24 current and voltage sensors, SHDSL modem, and DSLAMs (the SHDSL concentrator) are used in the network while sending PMU traffic to state estimator [27]. Figure 3 depicts the EPFL network topology.

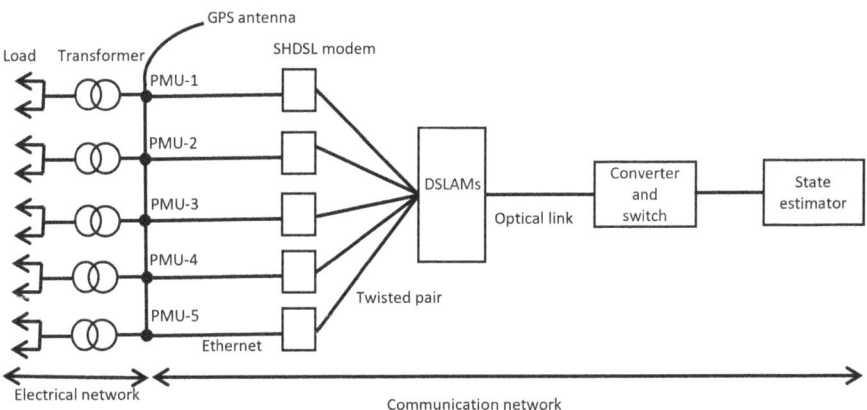

Fig. 3. EPFL PMU network topology

We do a plausibility check (B1) of the observed voltage and phase angle. Only plausible voltage and phase angle are sent from B1 to B2; otherwise, analyze the cause in B4. A final decision of the algorithms (sent from B2) is combined with the context in B3. We will instantiate an ontology for a power system in the context of the EPFL PMU network.

We investigate the root causes of an anomaly in the EPFL network. Suppose an anomaly is caused by physical changes such as the connection or disconnection of a power generator or the connection or disconnection of a load. In that case, it belongs to the real anomaly. In this case, the system needs some time to adopt the changes, so there is no mitigation.

If the anomaly is caused by failures in a PMU, a noisy or any sensor PMU, then it belongs to the non-malicious anomaly. In this case, the data is corrected by replacing the bad data. If there are no changes or problems in the electrical network that are causing an anomaly, then it can be a malicious anomaly caused by an attack. In this case, an alarm is generated. It covers blocks B3 and B4.

5 Conclusion and Future Work

We present a model for intrusion detection and trustworthy State Estimation in a Smart Grid. The proposed KBAIS uses anomaly detection algorithms, context, knowledge graph, and an ontology for anomaly detection. We discuss how we can combine algorithm-based anomaly detection and context-based anomaly detection. An ontology is proposed to identify anomalous behavior, its cause, and its appropriate mitigation strategy in the context of the EPFL PMU network. Our future endeavors will include reasoning, queries and information extraction; integrating the knowledge-based and data-based detection components; and implementing the proposed approach and using a rule-based method for the ontology approach to determine suitable appropriate mitigation strategy for any detected anomalies in the SG network to mitigate suspicious activities.

References

1. Paudel, S., Smith, P., Zseby, T.: Data integrity attacks in smart grid wide area monitoring. In: 4th International Symposium for ICS and SCADA Cyber Security Research (2016)
2. Paudel, S., Smith, P., Zseby, T.: Attack models for advanced persistent threats in smart grid wide area monitoring. In: Proceedings of the 2Nd Workshop on Cyber-Physical Security and Resilience in Smart Grids, CPSR-SG'17, pp. 61–66, New York, NY, USA. ACM (2017)
3. Lee, R.M., Assante, M.J., Conway, T.: Analysis of the Cyber Attack on the Ukrainian Power Grid. Technical report, SANS ICS and E-ISAC (2016)
4. Khediri, A., Laouar, M.R.: Prediction of breakdowns in smart grids: a novel approach. In: Proceedings of the International Conference on Computing for Engineering and Sciences, ICCES '17, pp. 82–85, New York, NY, USA. Association for Computing Machinery (2017)
5. McCarthy, et al.: Situation awareness for electric utilities. NIST Spec. Publ. **1800–7**, 02 (2017)
6. Alcaraz, C., Cazorla, L., Fernandez, G.: Risks and Security of Internet and Systems: 9th International Conference, CRiSIS 2014
7. Anwar, A., Mahmood, A.N.: Cyber Security of Smart Grid Infrastructure (2014)
8. Gokarn, V., Kulkarni, V., Saquib, Z.: Enhancing control system security of power grid using anomaly detection and behaviour analysis. In: 2017 International Conference on Advances in Computing, Communications and Informatics (ICACCI), pp. 1249–1255 (2017)
9. Paudel, S.: Detecting False Data Injection Attacks Against Smart Grid Wide Area Monitoring Systems. Ph.D. thesis, Technische Universität Wien (2021)
10. Lueckenga, J., Engel, D., Green, R.: Weighted vote algorithm combination technique for anomaly based smart grid intrusion detection systems. In: 2016 International Joint Conference on Neural Networks (IJCNN), pp. 2738–2742 (2016)

11. Alcaraz, C., Cazorla, L., Fernandez, G.: Context-awareness using anomaly-based detectors for smart grid domains. In: CRiSIS (2014)
12. Yoon, C., Dankel, D.D.: Domain specific knowledge-based information retrieval model using knowledge reduction. In: The Florida AI Research Society (2005)
13. Zhang, Z.Z., Hope, G.S., Malik, O.P.: Expert systems in electric power systems-a bibliographical survey. IEEE Trans. Power Syst. **4**(4), 1355–1362 (1989)
14. Russell, S.J., Norvig, P.: Artificial Intelligence - A Modern Approach, 3rd edn. Prentice Hall (2009)
15. Foley, S.N., Fitzgerald, W.M.: Decentralized semantic threat graphs. In: Cuppens-Boulahia, N., Cuppens, F., Garcia-Alfaro, J. (eds.) DBSec 2012. LNCS, vol. 7371, pp. 177–192. Springer, Heidelberg (2012). https://doi.org/10.1007/978-3-642-31540-4_14
16. European Committee for Electrotechnical Standardization CENELEC. Standard EN 50160 - Voltage Characteristics in Public Distribution Systems (2011)
17. Bergen, A.R., Vittal, V.: Power System Analysis, 2nd edn
18. Subba Rao, T.: A cumulative sum test for detecting change in time series. Int. J. Control **34**(2), 285–293 (1981)
19. Mohanty, S.R., Pradhan, A.K., Routray, A.: A cumulative sum-based fault detector for power system relaying application. IEEE Trans. Power Delivery **23**(1), 79–86 (2008)
20. Kullback, S., Leibler, R.A.: On information and sufficiency. Ann. Math. Stat. **22**(1), 79–86 (1951)
21. Kullback, S.: Information Theory and Statistics **22**(1), 79–86 (1968)
22. Paudel, S., Smith, P., Zseby, T.: Stealthy attacks on smart grid PMU state estimation. In: Proceedings of the 13th International Conference on Availability, Reliability and Security, ARES 2018, pp. 16:1–16:10, New York, NY, USA. ACM (2018)
23. Panda, M., Patra, M.: Ensemble voting system for anomaly based network intrusion detection. FULL PAPER Int. J. Recent Trends Eng. **2**, 01 (2009)
24. Mookiah, L., Dean, C., Eberle, W.: Graph-based anomaly detection on smart grid data. In: The Thirtieth International Flairs Conference (2017)
25. Dietterich, T.G.: Ensemble methods in machine learning. In: Kittler, J., Roli, F. (eds.) MCS 2000. LNCS, vol. 1857, pp. 1–15. Springer, Heidelberg (2000). https://doi.org/10.1007/3-540-45014-9_1
26. Brown, R.G., Hwang, P.Y.C.: Introduction to Random Signals and Applied Kalman Filtering with Matlab Exercises. John Wiley and Sons (2012)
27. Pignati, M., et al.: Real-time state estimation of the EPFL-campus medium-voltage grid by using PMUs. In: 2015 IEEE Power Energy Society Innovative Smart Grid Technologies Conference (ISGT), pp. 1–5 (2015)

AI System Engineering: Math, Modelling and Software

On the Solvability of the XOR Problem by Spiking Neural Networks

Bernhard A. Moser[1,2(✉)] and Michael Lunglmayr[2]

[1] Software Competence Center Hagenberg (SCCH), Hagenberg, Austria
bernhard.moser@scch.at
[2] Institute of Signal Processing, Johannes Kepler University, Linz, Austria
{bernhard.moser,michael.lunglmayr}@jku.at
https://www.jku.at/en/institute-of-signal-processing

Abstract. The linearly inseparable XOR problem and the related problem of representing binary logical gates is revisited from the point of view of temporal encoding and its solvability by spiking neural networks with minimal configurations of leaky integrate-and-fire (LIF) neurons. We use this problem as an example to study the effect of different hyper parameters such as information encoding, the number of hidden units in a fully connected reservoir, the choice of the leaky parameter and the reset mechanism in terms of reset-to-zero and reset-by-subtraction based on different refractory times. The distributions of the weight matrices give insight into the difficulty, respectively the probability, to find a solution. This leads to the observation that zero refractory time together with graded spikes and an adapted reset mechanism, reset-to-mod, makes it possible to realize sparse solutions of a minimal configuration with only two neurons in the hidden layer to resolve all binary logic gate constellations with XOR as a special case.

Keywords: Spiking Neural Networks (SNNs) · Leaky Integrate-and-Fire · Temporal Encoding · Reservoir Computing

1 Introduction

We consider the set $S = \{(0,0), (0,1), (1,0), (1,1)\}$ and all its binary partitions $P = [A, B]$, where $P = A \cup B$, $A \cap B = \emptyset$. XOR represents the special partition $P_{\text{XOR}} := [\{(0,0), (1,1)\}, \{(0,1), (1,0)\}]$. Solving the XOR problem refers to specifying a classification model that perfectly separates the subsets of the XOR partition P_{XOR}. Due to Radon's theorem [14] for any $d+2$ points in $\mathbb{R}^d$ there is a partition into two subsets with intersecting convex hulls. As a consequence, since for a linear classifier L with threshold ϑ the related pre-images $A := \{x \in \mathbb{R}^d : L(x) \geq \vartheta\}$ and $B := \{x \in \mathbb{R}^d : L(x) < \vartheta\}$ form a partition of convex sets, 4 points in $\mathbb{R}^2$ cannot be shattered by a linear classifier.

In this paper we study the problem under which conditions the set $S = \{(0,0), (0,1), (1,0), (1,1)\} \in \mathbb{R}^2$ can be shattered by a spiking neural network

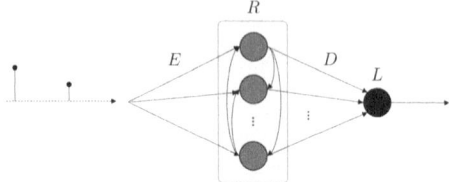

Fig. 1. Architecture of SNN considered in the paper, consisting of a reservoir R of fully connected hidden leaky integrate-and-fire (LIF) neurons of the same type (same threshold and leaky parameter) with weight matrix W and a linear output classifier L. The encoder weights $E = (1, 0, \ldots, 0)$ are fixed, W is generated randomly and the decoder weights D are learned.

SNN_W with a single hidden layer and weight matrix W. Figure 1 illustrates the architecture of such SNNs. We use the XOR problem as a vehicle to get insight into the effect of encoding, the choice of the reset mechanism and the difficulty to find a solution by considering the distribution of weights W. That is, we are looking for an as-simple-as-possible spiking neural network that allows to tune its weight matrix to realize any binary partition of interest.

The paper is structured as follows. Section 2 outlines related work on the XOR problem in the context of SNNs. Section 3 recalls the LIF model and the recently introduced reset-to-mod modification. Section 4 describes the setup of our experiments and discusses its results.

2 Related Work

Representing a simple non-linear problem that requires hidden units to transform the input into the desired output, the XOR problem is often considered a benchmark problem for testing neural network capabilities in solving more complex problems. This problem played an remarkable role in the early phase of AI. In their book [12], the neural network pioneers Marvin Minsky and Seymour Papert demonstrated that it is impossible for single-layer perceptrons (also referred to a first generation neural networks) to resolve the XOR problem. Unfortunately, incorrect citations in connection with these findings contributed to a significant decline in interest of neural network research in the 1970s, the so-called *AI winter*. It took another ten years before research in the field of neural networks began to take off in the 1980s [17]. In the meanwhile we encounter the so-called third generation of artificial neural networks in terms of spiking neural networks (SNNs) which are closer to the biological reference model by giving time a crucial role in information encoding and dynamics of the network [8,9].

The neurons in a spiking neural network (SNN) generate action potentials, or spikes, when the internal neuron state variable, called *membrane potential*, crosses a threshold. In contrast to conventional neural networks of the first and second generation, this way SNNs interconnect neurons that asynchronously process and transmit spatial-temporal information based on the occurrence of spikes that come from spatially distributed sensory input neurons [5,8].

Inspired from biology different information encoding principles with different characteristics have been proposed. Two main coding approaches can be distinguished for SNN-based systems: rate coding and temporal coding. For an overview, see, e.g. [2,8]. Rate coding aims to represent the intensity of a variable, e.g. voltage, by means of a spike frequency rate. This principle has been known in neurophysiology for many decades [1], so it has been experimentally discovered in most sensory systems such as the visual cortex and the motor cortex. However, rate coding comes also with drawbacks such as limitations due to slow information transfer and a long processing time. In contrast, temporal coding techniques use the precise timing of and between spikes to encode information. This includes the absolute timing with some reference, the relative timing of spikes triggered by different neurons, or simply the order in which neurons generate certain spikes.

The various information encoding variants are also taken up to tackle the XOR problem by spiking neural networks [3,4,6,11,16,18]. So, Bohte et al. [3] demonstrate a proof of concept of their SpikeProp algorithm by utilizing temporal encoding. While 0 is encoded with a *late* firing time and 1 is represented by *early* firing time. The related SNN topology consists of three input neurons (2 coding neurons and 1 reference neuron), 5 hidden neurons and a single output neuron. Due to convergence reasons this model does not allow a mix of both positive and negative weight. Therefore one of the hidden neurons is designed as an inhibitory neuron generating only negative sign spikes. In contrast, other authors such as [4,6,16,18] utilize rate encoding by representing 0 by spike trains of some frequency, e.g. 50 Hz, and 1 by another frequency, e.g., 100 Hz. By mimicking logic gates, in [18] the SNN topology for the XOR problem consists of two inputs, 2 hidden layers with 4 neurons each, and 2 output neurons, where the the first hidden layer is partially connected, based on neurons that are designed to respond on selected frequency ranges, resulting in two active neurons for any 0-1 combination. In the same spirit, also [16] mimics the functionality of logic gates but by utilizing additionally receptive fields between the LIF neurons to realize selective responses input frequencies. The resulting feed-forward SNN also consists of 2 input neurons and 4 LIF neurons in a hidden layer together with additional 2×8 receptive fields (RF) to filter the states $(0,0)$, $(0,1)$, $(1,0)$ and $(1,1)$, and two output neurons for 0 and 1, where the final decision is based on the winner-takes-all principle. [4] proposes four main layers of LIF neurons based on spike timing-dependent plasticity (STDP) as learning rule. While leaky integrate-and-fire (LIF) is the simplest neuron model for SNNs, also more advanced neuron models such as the Izhikevich neuron are used. Again using rate encoding, [6] studies the XOR problem by means of a feed-forward $2-2-1$ SNN architecture based on Izhikevich neurons. Its related 16 weights are found by genetic and evolutionary algorithms.

3 LIF Model and Preliminaries

The leaky-integrate-and-fire neuron model (LIF) with leaky parameter $\alpha > 0$ and threshold $\vartheta > 0$ uses integration which determines a recursive procedure to turn a signal f into a spike train $\eta(t) = \sum_k s_k \delta(t - t_k)$, where $s_k \in \mathbb{R}$ denotes the amplitude of the spike at time t_k. The time points t_k are recursively given by

$$t_{k+1} := \inf \left\{ T \geq t_k + t_r : \mathcal{T}\left[\int_{t_k}^{T} e^{-\alpha(t_{k+1}-t)} \big(f(t) + r_k \delta(t-t_k)\big) dt \right] \geq \vartheta \right\}, \tag{1}$$

where $\mathcal{T}[x] = x$ is either the identity (only positive threshold), or the modulus, $\mathcal{T}[x] = |x|$ (positive and threshold), $t_r \geq 0$ is the refractory time and $T = t_{k+1}$ is the first time point after t_k that causes the integral in (1) to violate the sub-threshold condition $|\int_{t_k}^{T} e^{-\alpha(t_{k+1}-t)} f(t) dt| < \vartheta$. The term $r_k \delta(t-t_k)$ refers to the reset of the membrane potential in the moment a spike has been triggered. In the standard definition of LIF for discrete spike trains, see [8], the reset is defined as the membrane potential that results from subtracting the threshold if the membrane's potential reaches the positive threshold level $+\vartheta$, or adding ϑ to the membrane's potential if a spike is triggered at the negative threshold level $-\vartheta$. In the case of bounded f the integral $g(t) := \int_{t_k}^{t} e^{-\alpha(t_{k+1}-t)} f(t) dt$ is changing continuously in t so that the threshold level in (1) is exactly hit. Consequently the resulting reset amounts to zero, i.e., $r_k = 0$ and the resulting amplitude s_k of the triggered spike is defined accordingly, i.e., $s_k = +\vartheta$, when the positive threshold value is reached, and $s_k = -\vartheta$ when the negative threshold value is reached. For a mathematical analysis and a discussion of how to define the reset r_k in the presence of Dirac impulses see [13].

Injecting weighted Dirac pulses the neuron's potential will show discontinuous jumps, and different reset variants are reasonable from an algorithmic point of view. Beyond the prevalent variants of *reset-to-zero* and *reset-by-subtraction*, see e.g. [7], recently we introduced *reset-to-mod* as a third option, see [13]. *reset-to-zero* means that the neuron's potential is reinitialized to zero after firing, while *reset-by-subtraction* subtracts the ϑ-potential u_ϑ from the membrane's potential that triggers the firing event. The third variant, *reset-to-mod*, can be understood as instantaneously cascaded application of *reset-by-subtraction* according to the factor n by which the membrane's potential u exceeds the threshold, i.e. $u = n\vartheta + r$, $r \in]-\vartheta, \vartheta[$. This means that *reset-to-mod* is the limit case of *reset-by-subtraction* with the refractory time t_r approaching to zero. In this case the residuum r results from a modulo computation and the amplitude of the triggered spike is set to $n\vartheta$.

As listed in Table 1, in total we get 6 LIF neuron model variants depending on the choice of thresholding (only positive, or positive and negative) and the reset variants *reset-to-mod*, *reset-by-subtraction* or *reset-to-zero*.

Table 1. LIF Spiking Neuron Model Variants

Model	Thresholding	Reset
Symmetric Reset-to-Mod (SRM)	positive and negative	reset-to-mod
Symmetric Reset-by-Sub (SRS)	positive and negative	reset-by-subtraction
Symmetric Reset-to-Zero (SRZ)	positive and negative	reset-to-zero
Positive Reset-to-Mod (PRM)	only positive	reset-to-mod
Positive Reset-by-Sub (PRS)	only positive	reset-by-subtraction
Positive Reset-to-Zero (PRZ)	only positive	reset-to-zero

4 Resolving Binary Logical Gates by SNNs

In contrast to the related work outlined in Sect. 2, our model consists only of a single input neuron and a single output neuron as illustrated in Fig. 1. The hidden layer is realized by a reservoir of N LIF neurons with randomly generated weights [15]. The decision is realized by a classical perceptron by summing up the weighted output spike trains ψ_k, $k = 1, \ldots, N$ at neuron L in Fig. 1. Note that the existence whether there is a solution or not can be checked by a linear program. If v_i^A denotes the vector of resulting sums for each output channel for the i-th input from the A class, and accordingly, v_j^B for j-th input from class B, then there the classes A and B can be separated linearly if and only if there is a vector D of decoder weights and a threshold ϑ such that for all i, j we have $\langle v_i^A, D \rangle \geq \vartheta$ and $\langle v_j^B, D \rangle < \vartheta$, where $\langle x, y \rangle = \sum_i x_i y_i$ denotes the standard inner product. This solvability problem can equivalently be decided by a homogenous problem with threshold 0 by checking the existence of $\widetilde{D}$ such that $\langle \widetilde{v}_i^A, \widetilde{D} \rangle \geq 0$ and $\langle \widetilde{v}_j^B, \widetilde{D} \rangle < 0$ for i and j, where $\widetilde{v}_i^A$ and $\widetilde{v}_j^B$ result from v_i^A, resp. v_j^B, by adding the additional coordinate -1. In turn, this problem can be equivalently decided by checking whether the set constituted of all $\widetilde{v}_i^A$ and $-\widetilde{v}_j^B$ can be linearly separated from the origin 0, i.e., whether the convex hull of the points $\widetilde{v}_i^A$ and $-\widetilde{v}_j^B$ contains 0 or not. This can be done by a linear program to solve for $x = (x_1, \ldots, x_n)$ satisfying $Px = 0$, $\sum_k x_k = 1$ and $x_k \geq 0$, where P is the matrix containing the points as column vectors, see, e.g., [10].

Our main objective is to investigate the effect of various hyper parameters on the distribution of solutions in the space of weights, that is how difficult it is to find a solution. Besides the XOR problem, we also consider all the other constellations of binary logical gates as illustrated in Fig. 2. For this we consider

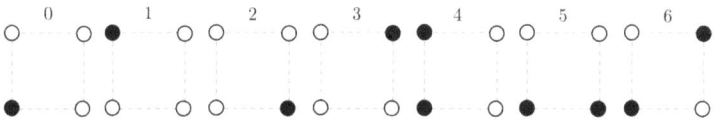

Fig. 2. Enumeration of all binary logical gates where XOR is represented by case 6.

temporal encoding in different constellations, also allowing mixed positive, negative spikes and spikes with different amplitudes (grades), see Fig. 3. The weights for the fully connected reservoir of LIF neurons are generated randomly based on uniform sampling the interval $[-1, 1]$ with discretization of 0.1. The probability evaluations are based on 100 runs in each considered constellation. Note that the variants A, A' and B of Fig. 3 represent 4 points in 2-dimensional space, whereas this is not the case for C. Therefore, only A or B are representation to which Radon's theorem of non-separability applies. Variant C circumvents the problem by increasing the dimensionality of the space from 2 to 4. Dimensionality enlargement by additional spikes might ease the problem as demonstrated in Table 5, but at the cost of sparseness. For $\beta = 1$ for all LIF variants SRM, SRS, SRZ, PRM, PRS, PRZ one can find solutions, where the probability to find a solution is greater the 10% for SRM and SRS. For $\beta = 0.5$ only for SRM and SRS there are solutions.

While encoding A does not work, the results for encoding variant B and leaky parameters $\beta = 1$, resp. $\beta = 0.5$, are shown in Table 2. Interestingly, our recently introduced *reset-to-mod* variant in terms of SRM (symmetric reset-to-mod) and PRM (positive reset-to-mod) gives the highest probability to find a solution, particularly for PRM with $\beta = 0.5$. Table 3 and Table 4 show the related mean and standard deviation of the l_1-norm of the output spike train. For PRM with $\beta = 0.5$ we obtain the sparsest solutions for all gates (Table 7).

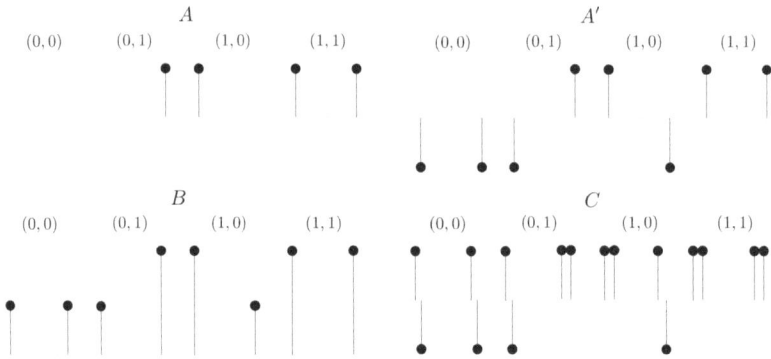

Fig. 3. Variants A, A', B and C for temporal encoding of logical gates by means of mixed positive and negative graded spikes.

Table 2. Solvability probability for encoding B, $E = (1,0)$, $\beta = 1$ (left), $\beta = 0.5$ (right)

Gate	SRM	SRS	SRZ	PRM	PRS	PRZ	Gate	SRM	SRS	SRZ	**PRM**	PRS	PRZ
0	30.0	27.5	0.0	49.5	49.5	0.0	0	63.5	22.5	0.0	**93.0**	25.0	0.0
1	2.5	2.5	0.0	1.0	1.0	0.0	1	20.0	0.0	0.0	**11.5**	0.0	0.0
2	3.0	2.5	0.0	1.0	1.5	0.0	2	14.5	0.0	0.0	**11.5**	0.0	0.0
3	29.5	28.5	0.0	50.0	49.5	0.0	3	70.0	22.5	0.0	**95.5**	25.0	0.0
4	1.5	3.0	0.0	1.0	5.0	0.0	4	10.5	2.0	0.0	**2.5**	1.0	0.0
5	4.0	4.5	0.0	0.5	1.5	0.0	5	5.0	2.0	0.0	**2.5**	1.0	0.0
6	2.0	1.0	0.0	1.0	0.5	0.0	6	7.5	0.0	0.0	**3.0**	0.0	0.0

Table 3. Mean of l_1-norm of output spike train of Table 2, $\beta = 1$ (left), $\beta = 0.5$ (right); "-" means not computable due to no spikes available.

Gate	SRM	SRS	SRZ	PRM	PRS	PRZ	Gate	SRM	SRS	SRZ	**PRM**	PRS	PRZ
0	12.7	12.8	-	4.4	6.6	-	0	5.9	5.2	-	**3.2**	4.9	-
1	32.2	37.0	-	5.0	18.0	-	1	24.0	-	-	**3.2**	-	-
2	41.0	15.6	-	5.0	15.0	-	2	11.4	-	-	**3.6**	-	-
3	5.6	10.2	-	3.4	4.5	-	3	3.1	3.5	-	**2.2**	3.4	-
4	31.7	50.3	-	6.0	16.2	-	4	42.0	6.5	-	**6.0**	7.0	-
5	17.0	27.3	-	6.0	16.0	-	5	7.3	6.5	-	**8.8**	7.0	-
6	18.8	36.5	-	6.0	8.0	-	6	3.8	-	-	**3.8**	-	-

Table 4. Standard deviation of l_1-norm w.r.t Table 3

Gate	SRM	SRS	SRZ	PRM	PRS	PRZ	Gate	SRM	SRS	SRZ	**PRM**	PRS	PRZ
0	24.4	21.3	-	4.3	5.5	-	0	16.6	1.8	-	**2.1**	1.6	-
1	47.9	39.6	-	1.0	1.0	-	1	57.7	-	-	**0.4**	-	-
2	47.4	4.1	-	1.0	4.3	-	2	33.9	-	-	**1.7**	-	-
3	9.2	25.0	-	3.5	3.7	-	3	1.5	0.9	-	**1.3**	0.8	-
4	36.3	45.6	-	0.0	3.4	-	4	75.1	0.9	-	**2.8**	1.0	-
5	25.1	31.8	-	0.0	7.0	-	5	3.2	0.9	-	**3.7**	1.0	-
6	22.7	28.5	-	1.0	0.0	-	6	1.0	-	-	**1.5**	-	-

Table 5. Solvability probability for encoding C, $E = (1,0)$, $\beta = 1$ (left), $\beta = 0.5$ (right)

Gate	SRM	SRS	SRZ	PRM	PRS	PRZ	Gate	**SRM**	**SRS**	SRZ	PRM	PRS	PRZ
0	61.0	72.0	66.0	41.0	41.0	40.0	0	**87.0**	**90.0**	88.0	39.0	39.0	39.0
1	19.0	30.0	27.0	3.0	3.0	6.0	1	**15.0**	**15.0**	14.0	5.0	5.0	5.0
2	24.0	29.0	22.0	7.0	7.0	7.0	2	**6.0**	**6.0**	6.0	0.0	0.0	0.0
3	57.0	66.0	70.0	59.0	59.0	40.0	3	**95.0**	**98.0**	95.0	44.0	44.0	44.0
4	34.0	41.0	27.0	62.0	61.0	38.0	4	**39.0**	**39.0**	37.0	44.0	44.0	44.0
5	14.0	17.0	8.0	5.0	5.0	3.0	5	**2.0**	**2.0**	0.0	0.0	0.0	0.0
6	13.0	17.0	8.0	6.0	6.0	3.0	6	**11.0**	**11.0**	8.0	5.0	5.0	5.0

Table 6. Mean of l_1-norm of output spike train of Table 5, $\beta = 1$ (left), $\beta = 0.5$ (right)

Gate	SRM	SRS	SRZ	PRM	PRS	PRZ	Gate	**SRM**	**SRS**	SRZ	PRM	PRS	PRZ
0	8.2	11.6	6.7	7.5	8.3	4.5	0	**3.7**	**3.6**	4.0	5.4	5.4	4.0
1	6.2	14.7	5.1	12.3	13.3	5.8	1	**7.7**	**7.7**	5.7	6.8	6.8	6.0
2	13.8	6.5	5.0	9.4	9.9	5.7	2	**8.0**	**8.0**	5.0	-	-	-
3	4.3	4.1	5.2	3.5	3.5	3.2	3	**3.5**	**3.5**	3.2	3.8	3.8	3.1
4	7.5	14.1	12.1	5.7	5.9	4.4	4	**6.2**	**6.2**	4.5	5.5	5.5	4.2
5	21.9	25.4	98.6	9.8	10.4	190.0	5	**12.0**	**12.0**	-	-	-	-
6	4.5	4.1	26.4	4.7	4.7	3.3	6	**5.1**	**5.1**	4.1	4.4	4.4	4.0

Table 7. Standard deviation of l_1-norm w.r.t Table 6

Gate	SRM	SRS	SRZ	PRM	PRS	PRZ	Gate	**SRM**	**SRS**	SRZ	PRM	PRS	PRZ
0	13.4	24.8	21.7	4.2	5.2	1.1	0	**2.5**	**2.5**	10.6	1.6	1.6	0.0
1	3.1	33.0	1.6	3.3	3.1	1.1	1	**3.0**	**3.0**	1.3	1.6	1.6	0.0
2	33.6	3.6	1.6	3.8	4.1	1.0	2	**3.5**	**3.5**	1.5	-	-	-
3	2.4	2.3	11.5	1.7	1.7	0.5	3	**1.0**	**1.0**	0.5	0.8	0.8	0.3
4	4.0	28.3	33.2	3.8	3.6	1.1	4	**2.3**	**2.3**	1.0	1.6	1.6	0.6
5	42.3	41.1	91.4	3.2	3.1	0.0	5	**2.0**	**2.0**	-	-	-	-
6	1.5	1.6	58.8	1.5	1.5	0.5	6	**1.4**	**1.4**	0.3	0.8	0.8	0.0

5 Conclusion

In this paper we study the problem to realize the decision problems of binary logical gates by means of spiking neural networks based on temporal encoding. It turns out that the choice of hyper parameters in terms of leaky parameter and the design of the reset mechanism in combination with the temporal encoding is crucial. In contrast to the standard setting of *reset-by-subtraction* we consider also *reset-to-mod* which can be understood as an instantaneous charge-discharge event with zero net voltage. It is shown that a temporal encoding of 0 and 1 based on graded spikes in combination with *reset-to-mod* and a reservoir of 2 fully interconnected LIF neurons provides the sparsest solution and that the weights can be found by uniform random initialization with a success rate of at least 3%, in our experiments 3 out of 100 runs. In future research we will also consider a LIF neuron as output layer which requires a generalization of the outlined solvability criterion based on the convex hull argument.

Acknowledgements. This work was supported (1) by the 'University SAL Labs' initiative of Silicon Austria Labs (SAL) and its Austrian partner universities for applied fundamental research for electronic based systems, (2) by Austrian ministries BMK, BMDW, and the State of Upper-Austria in the frame of SCCH and its project S3AI, part of the COMET Programme managed by FFG, and (3) by the *NeuroSoC* project funded under the Horizon Europe Grant Agreement number 101070634.

References

1. Adrian, E.D., Zotterman, Y.: The impulses produced by sensory nerve endings: Part 3. impulses set up by touch and pressure. J Physiol. **61**(4), 465–483 (1926)
2. Auge, D., Hille, J., Mueller, E., Knoll, A.: A survey of encoding techniques for signal processing in spiking neural networks. 53(6), 4693–4710 (2021)
3. Bohte, S.M., Kok, J.N., La Poutré, H.: Error-backpropagation in temporally encoded networks of spiking neurons. Neurocomputing **48**(1), 17–37 (2002)
4. Cyr, A., Thériault, F., Chartier, S.: Revisiting the XOR problem: a neurorobotic implementation. Neural Comput. Appl. **32**(14), 9965–9973 (2020)
5. Dayan, P., Abbott, L.F.: Theoretical Neuroscience: Computational and Mathematical Modeling of Neural Systems. The MIT Press (2001)
6. Enriqucz Gaytan, J., Gomez-Castaneda, F., Moreno-Cadenas, J.A., Flores-Nava, L.M.: Experimental spiking neural network: Solving the XOR paradigm with metaheuristics. In: 2018 15th International Conference on Electrical Engineering, Computing Science and Automatic Control (CCE), pp. 1–5 (2018)
7. Eshraghian, J.K., et al.: Training spiking neural networks using lessons from deep learning. arXiv (2021)
8. Gerstner, W., Kistler, W.M., Naud, R., Paninski, L.: Neuronal Dynamics: From Single Neurons to Networks and Models of Cognition. Cambridge University Press, USA (2014)
9. Maass, W.: Fast sigmoidal networks via spiking neurons. Neural Comput. **9**(2), 279–304 (1997)
10. Matouek, J., Gärtner, B.: Understanding and Using Linear Programming (Universitext). Springer-Verlag, Heidelberg (2006)
11. Matsumoto, K., Torikai, H., Sekiya, H.: Xor learning by spiking neural network with infrared communications. In: 2018 Asia-Pacific Signal and Information Processing Association Annual Summit and Conference (APSIPA ASC), pp. 1289–1292 (2018)
12. Minsky, M., Papert, S.: Perceptrons. MIT Press, Cambridge (1969)
13. Moser, B.A., Lunglmayr, M.: Spiking neural networks in the Alexiewicz topology: a new perspective on analysis and error bounds. arXiv preprint arXiv:2305.05772 (2023)

14. Radon, J.: Mengen konvexer Körper, die einen gemeinsamen Punkt enthalten. Math. Ann. **83**, 113–115 (1921)
15. Rahimi, A., Recht, B.: Weighted sums of random kitchen sinks: replacing minimization with randomization in learning. In: Koller, D., Schuurmans, D., Bengio, Y., Bottou, L. (eds.) Advances in Neural Information Processing Systems, vol. 21. Curran Associates, Inc. (2008)
16. Reljan-Delaney, M., Wall, J.A.: Solving the linearly inseparable XOR problem with spiking neural networks. In: 2017 Computing Conference, pp. 701–705 (2017)
17. Terrence, J.: Sejnowski. The Deep Learning Revolution. MIT Press, Cambridge, MA (2018)
18. Wade, J., McDaid, L., Santos, J.A., Sayers, H.: A biologically inspired training algorithm for spiking neural networks. In: Irish Signals and Systems Conference (ISSC), pp. 7–12, Derry, September 2007. Institution of Engineering and Technology

Risk Assessment in AI System Engineering: Experiences and Lessons Learned from a Practitioner's Perspective

Magdalena Fuchs[1(✉)], Lukas Fischer[1], Alessio Montuoro[1], Mohit Kumar[1], and Bernhard A. Moser[1,2]

[1] Software Competence Center Hagenberg (SCCH), Hagenberg, Austria
{magdalena.fuchs,lukas.fischer,alessio.montuoro,mohit.kumar, bernhard.moser}@scch.at
[2] Institute of Signal Processing, Johannes Kepler University, Linz, Austria

Abstract. Unlike the controlled conditions of AI system engineering laboratories, where adversarial vulnerabilities under specific threat models can be examined in isolation, in practical environments, such vulnerabilities are commonly intertwined with additional risks, including data or concept drift. In this paper, we explore the potential risks associated with the development and deployment of machine learning (ML) systems in real-world applications. We discuss secure ML engineering practices, their benefits, and their drawbacks and evaluate them based on their effectiveness in real-life use cases. Our study aims to provide a foundation for risk analysis and decision-making in practical ML applications where performance and security threats are highly intertwined.

Keywords: Federated Learning · Adversarial Vulnerability · Privacy Leakage · Data Drift · Concept Shift

1 Introduction

This paper examines the risk factors in developing and deploying machine learning (ML) systems, considering the divide between controlled environments and the complex, often conflicting requirements of practical ML applications. While performance and security threats can be isolated and analyzed separately in experimental settings, real-world use cases frequently reveal a theory-practice gap [12], as these risk factors are deeply interwoven, complicating their identification, analysis, and mitigation. Our investigation explores the nuanced challenges encountered in the practical engagement with these risks, discussing the extent to which they can be mitigated and the trade-offs to consider.

The research was supported by the Austrian ministries BMK and BMAW and the State of Upper Austria in the frame of the COMET Module Security and Safety for Shared Artificial Intelligence by Deep Model Design (S3AI) [(FFG grant no. 872172)], the AI Mission Austria Flagship Project FAIR-AI [(FFG grant no. 904624)] and the SCCH competence center INTEGRATE [(FFG grant no. 892418)].

© The Author(s), under exclusive license to Springer Nature Switzerland AG 2024
B. Moser et al. (Eds.): DEXA 2024 Workshops, CCIS 2169, pp. 67–76, 2024.
https://doi.org/10.1007/978-3-031-68302-2_6

We analyze risk factors that impact ML systems, both from theoretical and practical viewpoints and cover centralized and distributed ML systems. By examining prevalent secure ML engineering practices-such as homomorphic encryption, differential privacy, and federated learning-we assess their effectiveness across specific use cases. These cases, drawn from several domains, such as industry and healthcare, illustrate how the unique context of a ML application shapes its specific threat landscape.

The goal of this paper is to provide a foundation for structured risk analysis and decision-making regarding cost-efficacy compromises, guide research and development toward improving the robustness of ML systems against a variety of risks, and advocate for a holistic approach to safeguarding security and privacy in ML systems.

The paper is structured as follows: We first discuss common risk factors and considerations in productive ML systems. Then, we offer a theoretical perspective on ML threat models before presenting secure engineering concepts for ML systems. Next, we map out three specific ML scenarios and the risk factors most pertinent to them before describing specific real-life use cases and their requirements, providing context for the most relevant risk factors in each case. Finally, we discuss how ML engineering concepts can help mitigate the risks in each use case and the trade-offs that need to be considered.

2 Risk Factors, Entanglement and Side-Effects

Data-driven AI systems, such as ML models, and knowledge-based approaches, such as dynamic knowledge graphs, differ from the familiar standards of traditional software development. However, deviations from standardized engineering processes introduce new risk factors that fall into three categories: performance, security, and compliance issues. These can be organized into a pyramid of dependencies: The base consists of inherent technical risks associated with ML models' statistical and high-dimensional characteristics. Such fundamental issues can give rise to security and safety risks within specific application contexts. At the peak of the pyramid lie risks stemming from the interaction of technical and security risks with human factors and constraints imposed by regulatory policies, especially in AI applications involving human interaction.

- **Performance at Risk**: The performance of an AI system can degrade over time after the initial development for several reasons:
    - *Data and concept shift*: The characteristics of the inference data may gradually drift or abruptly change over time from the underlying training data, potentially causing the system to operate outside its intended range of validity [22]. Standard ML techniques rely on the i.i.d assumption that the data are sampled independently following the same distribution. However, this theoretical abstraction is rarely satisfied under practical circumstances, as in a federated learning setting where the data sources are separate [32].

- *Changing model due to updates at runtime*: The delivered system can change through a learning component, e.g., through active learning, by taking a human operator into the loop, or by employing updating mechanisms to feed the underlying knowledge model with actual data from a production process. However, such updates and new data can cause inconsistencies with previous states of the model [15].
- *Incorrect operation due to misleading interpretation in a human-AI setting*: If not explicitly addressed, data driven models often suffer from a lack of interpretability, leading to e.g. misconceptions and erroneous behavior of the system as a whole [3].
- **Security and Safety at Risk**: There are various threat scenarios that can jeopardize security and confidentiality through espionage or adversarial interference.
  - *Confidentiality and Privacy Leakage*: Involuntary data leakage is a phenomenon that accompanies ML and there are specific attacks on privacy explicitly designed to provoke such leaks, particularly during inference time, which is when publicly available models are most vulnerable [14]. For confidentiality issues in a federated learning setting see [32].
  - *System Integrity Vulnerability*: Integrity problems can arise when, for example, model training is delegated to an untrusted third-party service. In this situation, data poisoning may occur, which is an attack aimed at compromising the training data or influencing the learning process. For a detailed overview of poisoning attacks and their defenses in machine learning, see [5].
  - *Misleading Confidence Measures*: The performance issues mentioned above can also have an impact on safety as, for example, misleading confidence measures can provide a false sense of security and therefore can lead to incorrect judgments [13].
- **Compliance Rules to Consider**: Adhering to legal and regulatory standards is crucial, as non-compliance can expose ML systems to legal penalties and operational disruptions:
  - *Ethical Issues and Human Factor*: One of the biggest problems from an ethical point of view is the problem of oversight, i.e. the risk of people losing control. A fundamental question in this context is how to ensure correct behavior even in case of a possible malfunction or failure of an ML system.
  - *Increasing Compliance Policies and Standards*: There is a growing body of legal regulations and standardization guidelines concerning AI systems. The European AI Act [30] suggests a categorization into risk classes depending on whether humans are involved and the potential impact which complement or even overlap with other related legal policies such as the General Data Protection Regulation (GDPR) [10], the Data Act [9] and various ISO standards related to system quality.

2.1 A Theoretical Perspective

Privacy-Preserving ML. The goal of protecting sensitive information (that is embedded in training data) from any leakage through ML models has been addressed within the framework of differential privacy [1]. The classical approach for designing differentially private algorithms is output perturbation, where the idea is to perturb the function output via adding noise calibrated to the global sensitivity of the function [8]. The iterative nature of machine learning algorithms causes a high cumulative privacy loss, and thus, a high amount of noise needs to be added to compensate for the privacy loss. A moments accountant method [1], based on the properties of a privacy loss random variable, has been suggested to keep track of the privacy loss incurred by successive iterations for composition analysis. The moments accountant method can be combined with the use of the privacy amplification effect of subsampling to deal with iterative algorithms [24]. An obvious effect of adding noise into an algorithm for preserving differential privacy is the loss in the algorithm's accuracy. Therefore, efforts have been made to optimize the privacy-preserving noise-adding mechanism [16,17]. To mitigate the accuracy-loss issue of differential privacy, the post-processing property of differential privacy can be leveraged for fabricating new data samples by means of a geometrically inspired kernel machine [18].

Federated Learning. Federated learning has been a popular approach to collaborative learning among multiple parties without the exchange of raw data. However, the distributed nature of the data and their sampling from different local distributions pose challenges to the design of federated learning algorithms and their theoretical analysis. The issue of data heterogeneity in federated learning has been previously addressed by learning a personalized model for each client, assuming that data features share a common global representation, while statistical heterogeneity across clients is attributed to the labels [6]. The personalized federated learning problem has also been studied under the model-agnostic meta-learning framework with the goal of finding an initial shared model that can easily be adapted to local datasets by performing a few steps of gradient descent [11]. Another personalized approach is that clients, instead of fully utilizing the averaged global parameters for initialization, only select a subset of the global model's parameters and load the remaining parameters from previous local models [27]. Adversarial learning is another approach to deal with heterogeneous data features, where a discriminator is trained to distinguish the representations of the parties, while the parties aim to generate indistinguishable representations [21]. Alternatively, a clustered federated learning approach has been proposed based on the grouping of clients into clusters so that clients of the same cluster share the same model [25,29].

2.2 A Practical Perspective

This section explores essential ML engineering concepts for addressing the outlined risks to ML systems by enhancing privacy and security, including federated learning, homomorphic encryption, and differential privacy:

- *Federated Learning (FL)*: FL enables multiple devices to collaboratively learn a ML model without releasing their private data. A central server shares a model with all clients, which train the model on their local data. Despite alleviating some concerns regarding data sharing, FL remains vulnerable to certain threat models, such as membership inference attacks and data poisoning attacks [23].
- *Homomorphic Encryption (HE)*: HE allows a function to be evaluated over ciphertexts in such a way that the result of that evaluation, when decrypted, matches the result of the same operation carried out on the original plain texts. This allows computation directly on encrypted data without exposing the plain text data [31]. In ML, a model owner may therefore perform training or inference directly on encrypted data, ensuring only the original data owner is able to decrypt the result. However, HE comes at the expense of high computational costs and long training and inference times [19].
- *Differential Privacy (DP)*: DP is a mathematical formalism that guarantees an individual's privacy within a dataset by introducing randomness into the data release process, e.g., through additive noise. Despite the optimization of DP noise adding mechanism, there is a significant trade-off between privacy and utility: The noise introduced for privacy protection may impact the overall model performance.

From practical experience, we define three ML scenarios, illustrated in Fig. 1: *Scenario A* refers to a central ML setting where additive noise is used to achieve DP to reduce the threat of membership inference posed by malicious model users. *Scenario B* describes a central training with HE: A model trained on plaintext data runs inference on encrypted data, thus offering added security for the user's inference data. Finally, *Scenario C* shows a distributed FL setting, marking unique risks, such as the threat of data or model poisoning through malicious clients and the risk to model performance through data distribution shifts between clients.

Referencing these scenarios, three use cases with unique requirements are presented:

Use Case "Stress": A study aims to detect mental stress using heart rate measurements from a patient dataset of 3-minute-long R-R interval sequences and a stress score on a scale from 0 to 100. Since private data is directly contained within the data sequences, additional measures are required to guarantee the individual privacy of each patient's data. In order to ensure DP, a tailored noise-adding mechanism is used to achieve a given level of privacy loss. Privacy is the main concern in this use case pertaining to *Scenario A*.

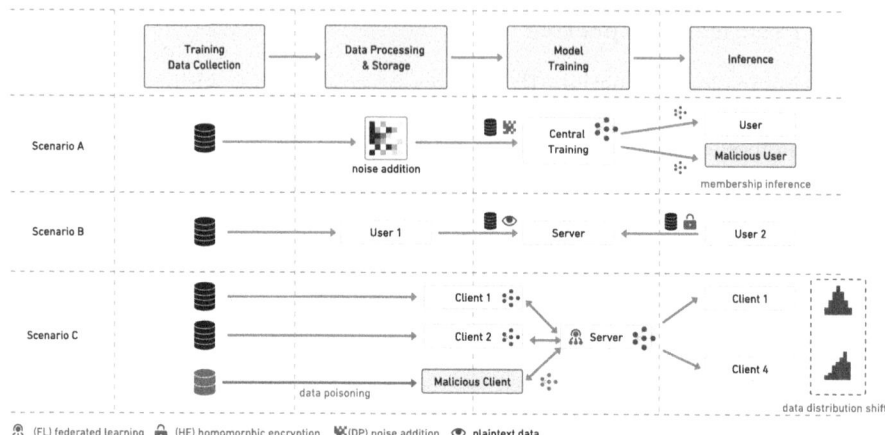

Fig. 1. Overview of considered ML scenarios and threats. Security and privacy threats are marked red and mitigation techniques are indicated in green. (Color figure online)

Use Case "Tools": A tool manufacturer provides a service to predict the wear of their tools, but customers are hesitant to share data. To solve this, a trusted customer can provide plaintext training features to train the prediction model, and standard users utilize the trained model for prediction by sending their homomorphically encrypted data to the inference server. The edge device used for data recording on drills is equipped with a Core i5-6442EQ processor, 16GB RAM, and a 480GB SSD and records spindle torque, speed, feed rate, and axis positions at 166 Hz. Standard users are provided with an interpretability score to determine how closely their inference data distribution matches the distribution of the training data. This use case is an example for *Scenario B*.

Use Case "Welding": A manufacturer of robotic welding machines plans to offer a service for their industrial customers that uses ML to predict welding outcomes and failures based on parameters such as welding angle, materials used, and machine settings. This data is recorded on the machines at a sampling frequency of either 250 Hz or 10 kHz. In order to profit from data diversity across different machines and use cases, multiple competing industrial actors who are distrustful of one another, collaborate to train a shared model in an FL framework to improve productivity and profitability. A main concern is non-detectability of training data shifts, since clients are competitors in this use case related to *Scenario C*.

Table 1 contains the main results of our structured risk analysis for the presented ML use cases. For each risk outlined at the beginning of Sect. 2, the susceptibility of the use case to that risk is evaluated on a scale of 0–3. Some threats can be generalized to a scenario, such as the risks depicted in Fig. 1, alongside some universal considerations for ML applications, like data distribution shifts and adherence to data privacy laws.

Table 1. Overview of the risks and issues encountered in each use case and the severity at which they were estimated for each use cases. Mitigation strategies are proposed for each considered risk, alongside a cost/benefit assessment for their implementation.

	Issues/Risks	Use Case "Welding" Relevance	Use Case "Welding" Notes / Requirements	Use Case "Tools" Relevance	Use Case "Tools" Notes / Requirements	Use Case "Stress" Relevance	Use Case "Stress" Notes / Requirements	Mitigation Strategy	Efficacy	Utility Decline	Cost/ Effort
Performance	non-availability of large amounts of training data		-	●●●	standard end user cannot provide training data	●	legal and ethical restrictions on sharing medical data	Data Augmentation	●○○	○○○	○○○
								Federated Learning	●●○	●○○	●●○
	training data distribution shifts	●●	training data shifts on one site affect model performance for all participants	●	-	●●	-	Anomaly Detection & Data Selection	●○○	○○○	●○○
	scalability and computational efficiency	●	-	●●●	inference must be possible < 1 min		-	Distributed Computing & Hardware Acceleration	●●●	○○○	●●●
Security & Privacy	bias of training data		-	●●●	inference data differs substantially from training data	●●●	bias in medical data can lead to misdiagnoses in undersampled groups	Federated Learning	●●○	●○○	●●○
	membership inference		participating parties are known	●●●	standard user does not provide training data	●●●	membership may implicate individuals in sensitive groups	Differential Privacy	●●●	●●○	○○○
								Federated Learning	●○○	●○○	●●○
								Homomorphic Encryption	●○○	●●●	●●●
	model inversion	●●●	temporal data distribution shifts should not be detectable by a collaborator	●●●	standard user does not provide training data	●●●	data is highly personal and sensitive	Differential Privacy	●●○	●●○	○○○
								Federated Learning	●○○	●○○	●●○
	data poisoning	●●●	risk of malicious participant	●●●	training data comes from trusted party only		-	Anomaly Detection & Data Selection	●○○	○○○	●○○
	adversarial attacks		-		-		-	Adversarial Training	●●○	○○○	●○○
Compliance	IP rights	●●●	training data is proprietary	●●●	model is proprietary			Federated Learning	●●○	●○○	●●○
								Homomorphic Encryption	●●●	●●●	●●●
	data privacy laws	●		●	-	●●●	use of personal and medical data	Differential Privacy	●●●	●●○	○○○
								Federated Learning	●●○	●○○	●●○
								Homomorphic Encryption	●●●	●●●	●●●
	explainability		-	●●●	end user is provided an explainability index	●●●	-	Federated Learning	●●○	●○○	●●○
								Explainable ML	●●○	○○○	●●○

●●● high/definite risk, action needed ●●● provable guarantees ●●● extreme added effort / severe performance decline
●● moderate risk, some action needed ●●○ strong improvement ●●○ considerable added effort / considerable performance decline
● slight risk, no urgent action needed ●○○ slight/possible improvement ●○○ possible or slight added effort / possible or slight performance decline
  not considered a risk ○○○ no positive effect ○○○ little to no added effort / no performance decline

However, Table 1 emphasizes that the specific context and requirements of a use case can shift the urgency with which certain threats need to be considered. The rightmost columns of Table 1 outline mitigation strategies for each risk, rating the strategy's *efficacy* in addressing the risk, as well as the associated *utility decline*, i.e., increased training times or reduced model accuracy, and the *cost* or *effort* required to implement the mitigation strategy. Ideal risk mitigation thus provides high enough efficacy to counter the severity with which the risk was assessed while keeping utility decline and cost/effort as low as possible.

In our analysis of risk mitigation strategies, we notably observe that FL falls short in countering privacy threats like model inversion and membership inference attacks. While FL may offer some protection in both cases compared to central ML scenarios, its effectiveness is not easily quantifiable and depends on the specifics of the training and the exact nature of the attack [2,7]. Considering this lack of privacy guarantees achieved through FL, implementing it specifically to counter privacy threats is likely not worth the associated cost and utility decline. FL is, however, effective in enhancing data availability and reducing data bias by integrating diverse datasets. Since no data sharing is required in FL systems, this can help adhere to data privacy laws and IP rights, though FL should not be considered a privacy guarantee by itself [4,28].

Additionally, we emphasize the expensive trade-off associated with HE: While HE provides strong guarantees in safeguarding against unauthorized disclosure of proprietary data, this comes at the compromise of severely impacted performance due to the computational complexity of operating on encrypted data [19]. Implementation of HE protocols into ML systems is also a complex task due to limitations in handling complex operations on encrypted data [20]. Considering these harsh trade-offs, their proportionality with respect to the severity of the risk assessment should be well-considered before implementing HE into an ML system. Table 1 also clarifies that HE is a security measure but cannot be considered a privacy measure since ML systems using HE remain vulnerable to model inversion attacks [26].

This detailed consideration of two techniques, FL and HE, provides an example of the intended use of our risk analysis and mitigation table and how it can aid in conducting cost-benefit analyses with respect to mitigation techniques for different threat scenarios.

3 Conclusion

In conclusion, methods that provide the strongest assurances of privacy and security, such as HE or DP, often come with significant trade-offs. Most risk mitigation techniques target threats to machine learning systems from a limited perspective, necessitating their combined application to achieve effective security. However, this can accumulate negative side effects, including diminished model accuracy and extended processing times. Our review of specific use cases underscores the fact that there is no one-size-fits-all approach to securing ML systems. This is due to the variety of threat models, security considerations, and performance aspects that must be taken into account. The priorities of threat mitigation thus need to be considered on a case-by-case basis, depending on the ML system architecture, the type and sensitivity of data used, and the performance requirements imposed upon the system.

This work offers a structured approach to analyzing security and privacy risks in ML use cases and provides a framework for determining suitable mitigation strategies without incurring unnecessary cost trade-offs. Moreover, it lays the foundation for future research, facilitating analyses of vulnerabilities in emerging ML architectures and aiding in developing robust security strategies for novel threat models.

References

1. Abadi, M., et al.: Deep learning with differential privacy. In: Proceedings of the 2016 ACM SIGSAC Conference on Computer and Communications Security, CCS 2016, pp. 308–318. ACM, New York (2016)
2. Abbasi Tadi, A., Dayal, S., Alhadidi, D., Mohammed, N.: Comparative analysis of membership inference attacks in federated and centralized learning. Information **14**(11) (2023)

3. Carvalho, D.V., Pereira, E.M., Cardoso, J.S.: Machine learning interpretability: a survey on methods and metrics. Electronics **8**(8) (2019)
4. Chalamala, S.R., Kummari, N.K., Singh, A.K., Saibewar, A., Chalavadi, K.M.: Federated learning to comply with data protection regulations. CSI Trans. ICT **10**(1), 47–60 (2022)
5. Cinà, A.E., et al.: Wild patterns reloaded: a survey of machine learning security against training data poisoning. ACM Comput. Surv. **55**(13s), July 2023
6. Collins, L., Hassani, H., Mokhtari, A., Shakkottai, S.: Exploiting shared representations for personalized federated learning. In: Meila, M., Zhang, T. (eds.) Proceedings of the 38th International Conference on Machine Learning. Proceedings of Machine Learning Research, vol. 139, pp. 2089–2099. PMLR, 18–24 July 2021
7. Dibbo, S.V.: Sok: Model inversion attack landscape: taxonomy, challenges, and future roadmap. In: 2023 IEEE 36th Computer Security Foundations Symposium (CSF), pp. 439–456 (2023)
8. Dwork, C., McSherry, F., Nissim, K., Smith, A.: Calibrating noise to sensitivity in private data analysis. In: Halevi, S., Rabin, T. (eds.) Theory of Cryptography, pp. 265–284. Springer, Heidelberg (2006)
9. European Commission, Directorate-General for Communication: Data Act - The path to the digital decade. Publications Office of the European Union (2022)
10. European Parliament, Council of the European Union: Regulation (EU) 2016/679 of the European Parliament and of the Council
11. Fallah, A., Mokhtari, A., Ozdaglar, A.: Personalized federated learning with theoretical guarantees: a model-agnostic meta-learning approach. In: Larochelle, H., Ranzato, M., Hadsell, R., Balcan, M., Lin, H. (eds.) Advances in Neural Information Processing Systems, vol. 33, pp. 3557–3568. Curran Associates, Inc. (2020)
12. Fischer, L., et al.: Ai system engineering-key challenges and lessons learned. Mach. Learn. Knowl. Extraction **3**(1), 56–83 (2021)
13. Guo, C., Pleiss, G., Sun, Y., Weinberger, K.Q.: On calibration of modern neural networks. In: Proceedings of the 34th International Conference on Machine Learning - Volume 70, ICML 2017, pp. 1321–1330. JMLR.org (2017)
14. Jegorova, M., et al.: Survey: leakage and privacy at inference time. IEEE Trans. Pattern Anal. Mach. Intell. **45**(7), 9090–9108 (2023)
15. Kemker, R., McClure, M., Abitino, A., Hayes, T.L., Kanan, C.: Measuring catastrophic forgetting in neural networks. In: Proceedings of the Thirty-Second AAAI Conference on Artificial Intelligence and Thirtieth Innovative Applications of Artificial Intelligence Conference and Eighth AAAI Symposium on Educational Advances in Artificial Intelligence. AAAI'18/IAAI'18/EAAI'18. AAAI Press (2018)
16. Kumar, M., Rossbory, M., Moser, B.A., Freudenthaler, B.: Deriving an optimal noise adding mechanism for privacy-preserving machine learning. In: Anderst-Kotsis, G., et al. (eds.) Proceedings of the 3rd International Workshop on Cyber-Security and Functional Safety in Cyber-Physical (IWCFS 2019), August 26-29, 2019, Linz, Austria, pp. 108–118. Springer, Cham (2019)
17. Kumar, M., Rossbory, M., Moser, B.A., Freudenthaler, B.: An optimal (ϵ, δ)−differentially private learning of distributed deep fuzzy models. Inf. Sci. **546**, 87–120 (2021)
18. Kumar, M., Moser, B.A., Fischer, L.: On mitigating the utility-loss in differentially private learning: a new perspective by a geometrically inspired kernel approach. J. Artif. Intell. Res. **79**, 515–567 (2024)

19. Lee, J.W., Kang, H., Lee, Y., Choi, W., Eom, J., Deryabin, M., Lee, E., Lee, J., Yoo, D., Kim, Y.S., No, J.S.: Privacy-preserving machine learning with fully homomorphic encryption for deep neural network. IEEE Access **10**, 30039–30054 (2022)
20. Li, J., Kuang, X., Lin, S., Ma, X., Tang, Y.: Privacy preservation for machine learning training and classification based on homomorphic encryption schemes. Inf. Sci. **526**, 166–179 (2020)
21. Li, Q., He, B., Song, D.: Adversarial collaborative learning on non-IID features. In: Krause, A., Brunskill, E., Cho, K., Engelhardt, B., Sabato, S., Scarlett, J. (eds.) Proceedings of the 40th International Conference on Machine Learning. Proceedings of Machine Learning Research, vol. 202, pp. 19504–19526. PMLR, 23–29 Jul 2023
22. Lu, J., Liu, A., Dong, F., Gu, F., Gama, J., Zhang, G.: Learning under concept drift: a review. IEEE Trans. Knowl. Data Eng. **31**(12), 2346–2363 (2019)
23. Lyu, L., Yu, H., Yang, Q.: Threats to Federated Learning: A Survey, March 2020. arXiv:2003.02133 [cs, stat]
24. Park, M., Foulds, J., Chaudhuri, K., Welling, M.: Variational bayes in private settings (vips). J. Artif. Intell. Res. **68**, 109–157 (2020)
25. Sattler, F., Müller, K.R., Samek, W.: Clustered federated learning: Model-agnostic distributed multitask optimization under privacy constraints. IEEE Trans. Neural Networks Learn. Syst. **32**(8), 3710–3722 (2021)
26. Shin, J., Choi, S.H., Choi, Y.H.: Is Homomorphic Encryption-Based Deep Learning Secure Enough? Sensors (Basel, Switzerland) **21**(23), 7806 (2021)
27. Sun, B., Huo, H., Yang, Y., Bai, B.: Partialfed: cross-domain personalized federated learning via partial initialization. In: Ranzato, M., Beygelzimer, A., Dauphin, Y., Liang, P., Vaughan, J.W. (eds.) Advances in Neural Information Processing Systems, vol. 34, pp. 23309–23320. Curran Associates, Inc. (2021)
28. Truong, N., Sun, K., Wang, S., Guitton, F., Guo, Y.: Privacy preservation in federated learning: an insightful survey from the GDPR perspective. Comput. Secur. **110**, 102402 (2021)
29. Vahidian, S., et al.: Efficient distribution similarity identification in clustered federated learning via principal angles between client data subspaces. In: Proceedings of the Thirty-Seventh AAAI Conference on Artificial Intelligence and Thirty-Fifth Conference on Innovative Applications of Artificial Intelligence and Thirteenth Symposium on Educational Advances in Artificial Intelligence. AAAI'23/IAAI'23/EAAI'23. AAAI Press (2023)
30. Walters, J., Dey, D., Bhaumik, D., Horsman, S.: Complying with the eu ai act (2023)
31. Yi, X., Paulet, R., Bertino, E.: Homomorphic encryption. In: Homomorphic Encryption and Applications, pp. 27–46. Springer, Cham (2014)
32. Zellinger, W., et al.: Beyond federated learning: On confidentiality-critical machine learning applications in industry. In: Proceedings of International Conference on Industry 4.0 and Smart Manufacturing (ISM) (November 2020), in press

From Paper to Pixels: A Multi-modal Approach to Understand and Digitize Assembly Drawings for Automated Systems

Raphael Seliger[✉], Sebnem Gül-Ficici, and Ulrich Göhner

Institute for Data-Optimized Manufacturing (IDF), University of Applied Sciences Kempten, Bahnhofstraße 61, 87435 Kempten, Germany
{raphael.seliger,sebnem.guel-ficici,ulrich.goehner}@hs-kempten.de

Abstract. The transition to Industry 4.0 intensifies the demand for advanced manufacturing techniques and efficient data processing capabilities. A notable challenge in engineering is that many older engineering drawings are only available in paper form, creating significant barriers for modern automated systems. This study tackles these challenges by employing advanced deep-learning techniques alongside traditional image processing to convert legacy engineering drawings into structured, machine-readable formats. Following this digitization process, this multimodal approach further processes drawings containing a lot of heterogeneous data by filtering non-essential details to isolate and extract critical features. This process enables the conversion of complex drawings into formats suitable for computer vision and deep learning applications. The structured datasets resulting from this process are then utilized to enhance the efficiency of automated processes significantly. For instance, they enable more efficient pick-and-place operations by providing the data necessary for machine learning-driven automation.

Keywords: Deep Learning · Computer Vision · Document Analysis · Engineering Drawing · Instance Segmentation

1 Introduction and Motivation

An engineering drawing (ED) visually represents shapes and concepts of systems such as mechanical systems, electrical circuits, or architectural structures [1]. These drawings are essential across various sectors, including oil and gas and mechanical engineering [2]. EDs act as a carrier of information that provides all necessary details for manufacturing, facilitating communication between different departments within a company and externally for outsourcing production to other manufacturing sites [3].

Understanding and digitizing these drawings automatically is becoming increasingly relevant in document analysis, particularly given that legacy engineering drawings and documents could represent a rich data source for many industries [2]. To analyze these drawings, it is often necessary to employ digital image processing techniques to understand text, symbols, tables, and the context among different elements [1]. Efforts to digitize these drawings into Computer-Aided Design (CAD) files have been ongoing since the 1980s and 1990s. However, this approach has gained more popularity in the scientific community due to the technical progress in computer vision in recent years [4]. Integrating drawings and diagrams into decision-making while simultaneously improving the efficiency of processes and the correctness of the files could represent a unique opportunity for optimization [1]. Nevertheless, despite significant improvements in image processing, particularly with the rise of neural networks, the automated analysis and processing of these EDs in Computer-Aided Manufacturing (CAM) still needs to be fully resolved [4].

2 Foundations and State of the Art

The following paragraphs discuss the potential applications of analyzing EDs in practical settings. They also explore various forms of EDs and their characteristic elements. Additionally, the current state of research is addressed, focusing on the automatic extraction and understanding of these elements.

2.1 Use Cases for Understanding and Digitizing EDs

In the introduction, particular attention was given to converting digital scans into CAD files. This process is motivated by several factors. Primarily, it enables the digitization of outdated technical drawings, many created before the rise of CAD-CAM solutions [6]. Additionally, some companies continue to exchange EDs in PDF format due to pre-agreed arrangements [5]. In many cases, no closed-loop CAD-CAM solutions are available in industrial brown-field environments. This lack of integration leads to unintended media disruptions, as EDs are frequently only available in paper form. These media discontinuities, while inefficient, cannot be directly eliminated due to the persistent nature of brown-field conditions [7]. In all these cases, the digitization of EDs represents a significant opportunity for optimization and efficiency improvements. However, Tombre notes that some large companies have already manually digitized their data inventory by now [8].

Nonetheless, there are additional use cases beyond digitization where understanding EDs automatically offers several advantages. Understanding the file's content and extracting a digital representation is helpful. This allows companies to utilize EDs alongside other data sources for data-driven analyses and decision-making [9]. For example, a searchable database of EDs for sparse parts can be established, or other computer vision systems can utilize the data for automation tasks such as pick-and-place operations. On the other hand, digitized EDs can also be used for quality control purposes to ensure the accuracy of these drawings [5]. Other potential applications could be the conversion of 2D data into 3D data [10], the automatic sorting and grouping of EDs for search operations or automatically determining manufacturing methods [4,28], and the automatic completion of drawings directly in the CAD software [29]. All these processes significantly improve the efficiency and quality of data management in companies that rely on experts' manual and time-consuming work for conversion, creation, and quality control [5].

2.2 Types and Characteristics of EDs

In mechanical engineering, various types of EDs represent a product, depending on the type of information and the country of origin. The most frequently used types are assembly drawings and production drawings. Assembly drawings provide full or partial details on the assembly process for various components, and production drawings focus on individual parts, including detailed specifications necessary for manufacturing, such as surface finishing, tolerances, and grading [6]. In addition to these drawings, other sectors also utilize specific EDs. For example, the architecture industry uses architectural drawings, while the process industries often rely on piping and instrumentation diagrams (P&IDs). Other drawings are chemical process drawings, complex circuit drawings, process flow diagrams, and sensor diagrams [1].

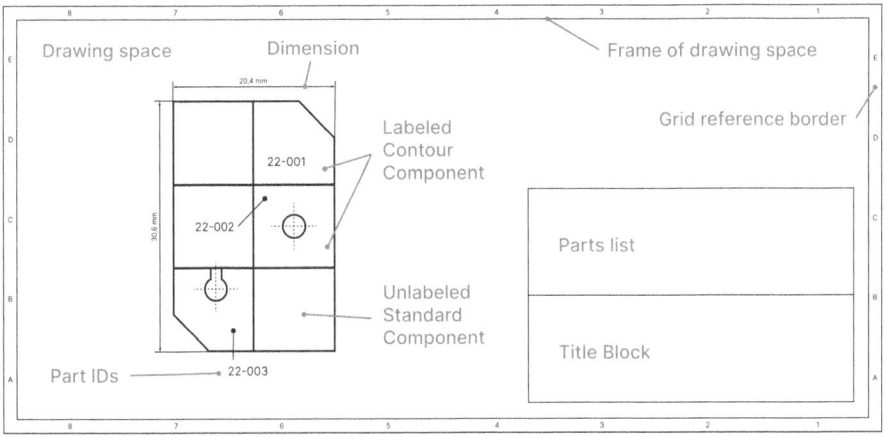

Fig. 1. This figure illustrates a typical assembly drawing employed in pick-and-place processes alongside common elements.

The elements of professional production EDs are standardized by organizations such as the International Organization for Standardization (ISO) and the American Society of Mechanical Engineers (ASME). These standards help to ensure a universal understanding of a product's geometric requirements across the industry [6]. Common elements are visualized in Fig. 1, including a title block and parts list table at the bottom or side, providing essential information about the assembly components. Surrounding the central drawing space is a grid reference border, which aids in locating different parts within the drawing. The main drawing space is framed to distinguish the detailed assembly area, where various parts are depicted along with their corresponding part IDs and dimensions. Each component is precisely drawn to scale, showing the interconnections and layout necessary for the assembly process. In other EDs, symbols and special characters are frequently used to visualize relationships or communicate specific logic [6]. Many elements in EDs are standardized, but specialized syntaxes, notations, and symbols can vary, reflecting company-specific or industry-specific practices. Recognizing and adapting to these variations is crucial for accurate interpretation [5].

2.3 Challenges of EDs in Computer Vision

Processing and analyzing EDs is quite similar to standard image-processing tasks, which allows using various typical image pre-processing and analysis techniques from other areas of document analysis. However, due to the complexity of these drawings, achieving a fully automated framework for the reading, processing, and analysis still needs to be resolved [11]. Research in digitizing EDs has been ongoing since the late 1980s [12] and yet remains challenging. Typically, EDs still require manual inspection, digitization, or creation in production processes [4]. Existing approaches have attempted to extract data from raster images in different aspects, for example, for understanding or converting scans into CAD files [8,13]. Before 2010, most researchers attempted text and graphic segmentation using traditional computer vision techniques [5]. Subsequent efforts explored using convolutional neural networks (CNNs) or hybrid approaches to enhance segmentation [1,2,14,15].

Table Recognition. Table recognition and tabular data extraction in EDs are often vital for digitalization efforts. The process typically begins with the detection of the table, followed by categorization, extracting text using an optical character recognition (OCR) system, and applying an ordering text [6] or table structure recognition algorithm to process the textual information from tabular formats appropriately. Despite the availability of numerous methods for detecting tables in document images and breaking them down into their structural components, these tasks still need to be completed for modern document processing systems [19].

Optical Character Recognition. In the initial stages of digitizing EDs, OCR was achieved through various methods such as identifying arrowheads to locate text, eliminating noise and graphical elements to isolate text, or employing primary neural networks for text recognition on raster images [8,16,17]. However, these initial methods needed to be more robust and efficient for reliable analysis of EDs [6]. Therefore, deep learning approaches are nowadays used for OCR recognition, which must recognize both standard letters and special symbols [8,18]. Typically, OCR systems involve several key steps: image acquisition, text detection, and text recognition. Additional processes often include image preprocessing, data augmentation, segmentation, and postprocessing to enhance accuracy [1]. For this application, Tesseract-OCR is often employed for recognizing standard letters, or Keras-OCR is frequently used when the alphabet needs to be expanded with additional characters [6].

Recent advances in OCR technology besides EDs include transformer-based approaches such as DocOWL [33] and Idefics2 [34]. These approaches have shown significant progress in processing text and image data. These models utilize transformers to extract complex relationships and semantic meaning from large datasets, making them promising for use in EDs. While traditional methods like CNNs and Recurrent Neural Networks (RNNs) have text and symbol recognition limitations, transformer models offer enhanced capabilities in capturing contextual information and processing complex layouts [35]. For example, DocOWL can be used not only to recognize text but also to understand the semantic structure and relationships between different elements [33], potentially improving the accuracy and efficiency of ED digitization. These advances represent a promising direction for future research and applications.

Symbol Detection. For some EDs, the recognition of special symbols is vital. Initial efforts in classifying symbols in EDs relied on comparisons with a series of symbols [4]. Recent studies frequently use CNNs for detection, representation, and classification [1,2,26]. Often, the focus in the research is on P&IDs, which present significant challenges in symbol classification and analysis due to their complexity and the number of symbols they contain [1,5,14,15,27]. On average, a drawing can feature around 100 shapes, including symbols, connectors, and text. These diagrams incorporate various special symbols, which differ across industries, and a network of connecting lines that depict physical and logical relationships between symbols [1]. The automatic analysis of symbols helps in verifying the accuracy of diagrams and facilitates their digitization [5].

Data Augmentation. The challenge of recognizing text, symbols, and other elements in technical drawings is intensified because these drawings often contain sensitive company information and are not publicly available. This lack of accessibility means there are no freely available datasets for training models [7]. Additionally, when a company provides data, acquiring labeled data for such drawings poses a significant challenge because it demands the involvement of individuals with the necessary technical expertise to participate in the labeling process [25]. However, using a generator to expand the dataset size artificially can significantly improve the model's performance [7]. This method involves separating dimensional elements from object contour lines in the initial files, randomly generating new dimensional configurations, and rasterizing these to produce a diverse set of labeled training data [7,25].

For data augmentation, applying random variations in sharpness, contrast, and brightness to simulate different image qualities found in actual drawings can improve model robustness [7]. Models trained with synthetic data outperform those trained with limited actual drawings [7]. Although capable of generating numerous synthetic drawings, the limited number of original actual drawings restricts the diversity, as the contour shapes remain the same. This constraint results in limiting the variety of synthetic data [25]. The accuracy of the generated data heavily depends on the quality of the generator and preprocessing methods. Weaknesses in these areas can impact overall performance [25].

Generating synthetic data can improve the generalizability of models by providing more training data, which helps to avoid overfitting [25]. However, while models trained with synthetic data perform well on similar synthetic data, their generalizability to entirely new, unseen actual drawings still needs to be improved [7]. The overall accuracy of the method is constrained by the precision of the preprocessing steps, such as vectorization, with any weaknesses in these steps potentially affecting the model's overall performance. Synthetic drawings are based on existing data structures and thus remain limited in their diversity. Complete automation of the drawing synthesis process is challenging due to the need for manual adjustments in CAD software [25].

Vectorisation and Classification. Previous studies have explored the vectorization of EDs from raster images to improve data retrieval and manipulation. The vectorization process often relies on line detection techniques like the Hough transform, which, although widely used, can struggle by incorrectly identifying line segments or producing parallel line pairs [4]. A neural network-based contour tracing technique that improves the accuracy of line segment detection can address these limitations [20] by enhancing the tracing of complex shapes, such as the crosses indicative of circular holes typically found in EDs [4]. Other approaches focus on reconstructing primitive geometric shapes, such as polygons, using various features for shape recognition [21] or methods like transformer-based networks for reconstructing line drawings [22]. Research into the classification of engineering drawings has evolved to include advanced machine learning models to address the challenges of processing high-resolution images [4]. A classification approach often involves the usage of graph-matching techniques to cluster vector EDs based on similarities, although scaling these techniques for high-resolution images remains a challenge [23,24]. An improvement to these approaches is using graph neural networks, which combine graph matching principles, image segmentation, and line-tracing to enable more accurate classification of EDs and enhance retrieval and analysis capabilities [4].

Conclusion. Currently, no open-source solution can reliably analyze, digitize, or understand EDs. While some aspects have been explored within certain types of EDs, comprehensive solutions still need to be developed. Therefore, significant research gaps still exist in this area. Besides academic research, commercial products like Werk24 offer feature extraction from EDs. However, these products typically have closed APIs, restricting access to their data and the methodologies employed. This limitation prevents the scientific community's ability to enhance and build upon these technologies [6].

3 Methodology and Project Objectives

This research focuses on digitizing assembly diagrams, which have been relatively understudied compared to other types of EDs. Assembly diagrams are distinguished mainly by the critical need to recognize individual components and sometimes their precise location. This last characteristic is particularly relevant for the automation of pick-and-place processes, where accurate component placement is essential. The study aims to close the research gaps in digitizing assembly diagrams by focusing on this aspect. A multi-modal approach is employed to tackle this project objective, combining deep learning for computer vision with traditional computer vision techniques. The approach is both theoretical and practical, involving the direct application of methodologies in a real-world case study. Action Research was chosen as a methodological framework, which facilitates iterative exploration and practical intervention.

4 Design and Development

First, a comprehensive preprocessing framework and various methods were developed to ensure data integrity and consistent format-independent processing across all steps. Classes were developed for uniformly handling images, vectors, bounding boxes, segmentation masks, and objects. For preprocessing, PDF data was converted into images by choosing the appropriate Dots per Inch (DPI) setting to enable segmentation. Too low resolution would yield segmentation masks of poor quality, while too high resolution would result in large images.

4.1 Segmentation and Sliding Window Approach

Our research shows that working with EDs poses unique challenges. This is mainly because no pre-trained models are designed explicitly for EDs or similar technical drawings. Another significant issue is the need for publicly available training datasets, as companies keep most relevant data confidential. We had a small dataset consisting of only 15 real-world assembly drawings. The limited amount of data restricted the number of test plans we could conduct and required us to adjust our approach. This scarcity of data is a common challenge in this field, as indicated by our literature review.

We considered using synthetic data to increase the limited training data available. However, synthetic data augmentation has drawbacks, especially for assembly diagrams. Current approaches often focus on using existing components with altered text, which does not meet our need for precise placement and contour accuracy. Furthermore, synthetic data can introduce biases and may only partially capture the complexity of real-world data, making it less effective for our specific requirements. Therefore, we aim to evaluate the relatively new Segment Anything (SAM) model [30] as an alternative. This involves leveraging its zero-shot capabilities to recognize objects from various classes without specific training data. This approach helps seamlessly integrate new symbols and configurations into the system without needing a proprietary model or specific datasets. Its "Everything" mode evaluates and scores the segmentation quality using a dense grid of points to generate masks. The performance of the SAM model is highly dependent on its hyperparameters. The 'threshold' parameter was set to 0.5, ensuring the model neither over-segments nor misses critical parts. SAM segments irrelevant elements deemed non-essential, and approximately 65% of the recognized objects were discarded through a filtering process, while the remaining 35% were further processed.

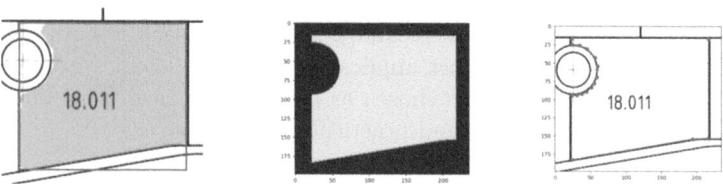

Fig. 2. This figure illustrates from left to right first the SAM mask, then the post processed mask and finally the extracted contours.

4.2 Classification and Postprocessing

Next, the resulting SAM masks need to be classified. For this purpose, we have chosen ResNet18 [31], with a final layer customized to a size of 5, and utilized it to classify arrows, cutouts, and parts in the assembly drawing masks. We used transfer learning with fine-tuning to train the pre-trained ResNet18 model on our dataset of ED masks. This approach helped us adapt the model to our specific use case, requiring less data and computational resources than training a model from scratch.

To ensure generalizability, we performed k-fold cross-validation with k=5, dividing the dataset into five equally sized subsets. The model was trained on four subsets and validated on the fifth. This process was repeated five times, using each subset once as a validation dataset. The cross-validation results show consistent accuracy across the different subsets of the dataset, indicating good generalizability of the model. The average accuracy across the five folds was 97.1%, with a standard deviation of 1.2%. The final ResNet18 was trained on 50 epochs and 12,220 annotated images resulting from the SAM masks combined with the actual drawing in different channels, achieving over 98% accuracy. An 80% to 20% split was used for training and testing, ensuring a robust and generalizable model. The dataset contains the classes cutout, standard contour, contour with arrow, contour with text, text, and arrow. For the ResNet18 model, key hyperparameters included the batch size of 32, and the Adam optimizer for its adaptive learning rate capabilities, facilitating faster convergence with a learning rate initially set at 0.001 and adjusted using learning rate scheduling.

In the literature on EDs, TesseractOCR has been predominantly used. Although recent advancements with transformer models and OCR are promising, we initially limited our approach to the well-established TesseractOCR, focusing on extracting contour information within the context of pick-and-place processes. Figure 2 shows an example mask with applied postprocessing. OpenCV's 'findContours' and 'approxPolyDP' functions were utilized for shape recognition. Additionally, the extracted objects were combined to represent the original layout coherently.

4.3 Dimensions, Positioning and Assignment of Part IDs

The dimensioning process scaled pixel values to real-world millimeters using a conversion factor derived from standard parts in each plan, which have consistent measurements across different drawings. Assigning part IDs in EDs is critical and involves challenges like handling complex cases where arrows contain contextual information about the connected components. Figure 1 shows the three types of parts: some are not labeled, some are labeled, and some are only labeled through an arrow. The initial method used arrows identified by the SAM model to locate nearby text and components, but this proved too simplistic, especially with closely situated arrow-marked components. Therefore, a more refined method was developed, which involves localizing the arrow and its head using Faster-R-CNN [32], determining the arrow's direction from the bounding box's aspect ratio and arrowhead position, calculating the arrow's angle for precise directional extrapolation, and extrapolating in the arrow's direction until it intersects with a text segment. After extracting dimensions and positions and assigning part IDs, a final postprocessing step involved duplication detection. During the Faster-R-CNN training, we set the learning rate to 0.005 and the batch size to 2. The momentum was set to 0.9 to accelerate the Stochastic Gradient Descent (SGD) process, and the weight decay was set to 0.0005 to regularize the model and prevent overfitting.

5 Results and Discussion

We developed a prototype system to test the accuracy and effectiveness of our methods. The system included a backend with an API, a database, and a frontend application for interactive testing of real-world assembly drawings. To evaluate the effectiveness of our approach, we tested it on a set of 15 actual EDs. We used ten drawings for training and reserved five for validation. The results were promising, with the system achieving a 98% accuracy rate on the validation set for the extracted contours.

The approach shows promise, but it also has some limitations. The chosen TesseractOCR has trouble recognizing vertical text and detecting text near contour lines. Another challenge is the SAM model's high computational intensity, which may not be practical for large-scale industrial applications in this context. The proposed approach requires significant computational resources. Smaller documents can be segmented in 5 to 10 min. However, larger ones require much memory (100 dpi, 32GB RAM) and must be segmented into overlapping sliding windows, taking up to 30 min for digitization. A custom model trained on synthetic data could provide a more efficient solution.

In addition to the current neural network solution, a hybrid approach, in combination with traditional computer vision techniques like Blob Coloring, could speed up processing without compromising robustness. Additionally, some improvements regarding data handling could be explored, like storing contours as continuous polygonal chains in formats like DXF or SVG, which would streamline operations and improve integration. Besides that, a future version could incorporate more elements like outer dimensions and parts lists, refining the system's accuracy and utility. Furthermore, a synthetic dataset is suggested to address the lack of available assembly drawings to train a potential transformer model for text recognition and a better classification and segmentation approach.

The ResNet18 model used pre-trained weights from ImageNet and was fine-tuned to effectively generalize to new, unseen assembly drawings, maintaining high accuracy and robustness in practical applications. The proposed arrow detection and direction approach performed well for the evaluated assembly drawings, with no issues identified. However, it is essential to note that the current approach was validated using only five assembly diagrams. While the accuracy was high, this limited sample size may not cover all potential variations in real-world assembly drawings. Future work should include a more extensive dataset to ensure robustness.

6 Conclusion and Outlook

Overall, the prototype's high accuracy in extracting the components alongside their positions highlights its potential but also identifies challenges and opportunities for future system enhancements. This study presented a robust method for automating the digitalization of elements from assembly drawings, improving the efficiency of the digitization process. The system can be used to extract a foundational database from legacy assembly drawings to automate pick-and-place operations, and the findings can also be transferred to other EDs, especially the model selection from state-of-the-art advancements in computer vision.

However, several improvements can be made to address computational demands. The current SAM segmentation process takes up over 75% of the total processing time, leaving room for improvements. Future work should refine the segmentation step, incorporating the point mode of SAM for quick validation and correction. Other solutions could be developing smaller, domain-specific models with fewer parameters or combining traditional approaches with deep learning enhancements in a hybrid system architecture. A transformer-based OCR approach could be a promising direction for future research and applications for improving text recognition.

Acknowledgements. Some of the work described in this paper was part of a semester project by computer science students at the University of Applied Sciences Kempten. The authors especially want to thank Samuel Pfalzer and Raphael Gut, who made many valuable contributions to this project.

Disclosure of Interests. The authors have no competing interests to declare that are relevant to the content of this article.

References

1. Moreno-García, C.F., Elyan, E., Jayne, C.: New trends on digitisation of complex engineering drawings. Neural Comput. Appl. **6**(31), 1695–1712 (2019)
2. Elyan, E., Jamieson, L., Ali-Gombe, A.: Deep learning for symbols detection and classification in engineering drawings. Neural Netw. **129**, 91–102 (2020)
3. Labisch, S., Wählisch, G.: Technisches Zeichnen. Eigenständig lernen und effektiv üben. 6th edn. Springer Vieweg Wiesbaden, Wiesbaden, Germany (2020)
4. Xie, L., Lu, Y., Furuhata, T., et al.: Graph neural network-enabled manufacturing method classification from engineering drawings. Comput. Ind. **142**, 103697–103707 (2022)
5. Dzhusupova, R., Banotra, R., Bosch, J., et al.: Using artificial intelligence to find design errors in the engineering drawings. J. Softw. Evol. Process **35**(12), 1–19 (2023)
6. Toro, J.V., Wiberg, A., Tarkian, M.: Optical character recognition on engineering drawings to achieve automation in production quality control. ASME International Design Engineering Technical Conferences and Computers and Information in Engineering Conference **3**, 1–19 (2023)
7. Schlagenhauf, T., Netzer, M., Hillinger, J.: Text Detection on Technical Drawings for the Digitization of Brown-field Processes. Procedia CIRP **118**, 372–377 (2023)
8. Tombre, K.: Analysis of engineering drawings: State of the art and challenges. In: Handbook of Character Recognition and Document Image Analysis, vol. 1389, pp. 257–264 (1998)
9. Van Daele, D., Decleyre, N., Dubois, H., et al.: An automated engineering assistant: learning parsers for technical drawings. In: Proceedings of the AAAI Conference on Artificial Intelligence 35(17), pp. 15195–15203 (2021)
10. Kanungo, T., Haralick, R.M., Dori, D.: Understanding engineering drawings: a survey. Proceedings of the 1st International Conference on Graphics Recognition 1, pp. 119–130 (1995)
11. Elyan, E., Moreno-García, C.F., Jayne, C.: Symbols classification in engineering drawings. In: International Joint Conference on Neural Networks, pp. 1–8 (2018)
12. Krause, F.L., Jansen, H., et al.: Automatic scanning and interpretation of engineering drawings for CAD-Processes. CIRP Ann. **38**(1), 437–441 (1989)
13. Lu, T., Yang, Y., Yang, R. et al.: Knowledge extraction from structured engineering drawings. In: Fifth International Conference on Fuzzy Systems and Knowledge Discovery, pp. 415–419 (2008)
14. Mani, S., Haddad, M., Constantini, D., et al.: Automatic digitization of engineering diagrams using deep learning and graph search. In: IEEE/CVF Conference on Computer Vision and Pattern Recognition Workshops, pp. 673–679 (2020)
15. Kang, S., Lee, E., Baek, H.: A digitization and conversion tool for imaged drawings to intelligent piping and instrumentation diagrams (P & ID). Energies **12**(13), 1–26 (2019)

16. Das, A.K., Langrana, N.A.: Recognition of dimension sets and integration with vectorized engineering drawings. In: Proceedings of 3rd International Conference on Document Analysis and Recognition 1, pp. 347–350 (1995)
17. Lu, Z.: Detection of text regions from digital engineering drawings. IEEE Trans. Pattern Anal. Mach. Intell. **20**(4), 431–439 (1998)
18. Haar, C., Kim, H., Koberg, L.: AI-based engineering and production drawing information extraction. In: Flexible Automation and Intelligent Manufacturing: The Human-Data-Technology Nexus, pp. 374–382 (2023)
19. Schreiber, S., Agne, S., Wolf, I. et al.: DeepDeSRT: deep learning for detection and structure recognition of tables in document images. In: 14th IAPR International Conference on Document Analysis and Recognition (ICDAR), pp. 1162–1167 (2017)
20. Bessmeltsev, M., Solomon, J.: Vectorization of line drawings via polyvector fields. ACM Trans. Graph. **38**(1), 1–12 (2019)
21. Mane, A., Adhikari, R., Gadgil, S. et al.: Investigating application of machine learning in identification of polygon shapes for recognition of mechanical engineering drawings. In: International Conference on Nascent Technologies in Engineering (ICNTE), pp. 1–6 (2019)
22. Egiazarian, V., Voynov, O., Artomov, A., et al.: Deep vectorization of technical drawings. Computer Vision - ECCV, pp. 582–598 (2020)
23. Fonseca, M.J., Ferreira, A., Jorge, J.A.: Content-based retrieval of technical drawings. Int. J. Comput. Appl. Technol. **23** (2/3/4), 86-100 (2005)
24. Kuchuganov, V.N., Kuchuganov, A.V., Kasimov, D.R.: Clustering algorithm for a set of machine parts on the basis of engineering drawings. Program. Comput. Softw. **46**(1), 25–34 (2020)
25. Zhang, W., Chen, Q., Koz, C., et al.: Data augmentation of engineering drawings for data-driven component segmentation. In: Volume 3A: 48th Design Automation Conference (DAC), pp. 1–12 (2022)
26. Fu, L., Kara, L.B.: From engineering diagrams to engineering models: visual recognition and applications. Comput. Aided Des. **43**(3), 278–292 (2011)
27. Jamieson, L., Moreno-García, C.F., Elyan, E.: Deep learning for text detection and recognition in complex engineering diagrams. In: International Joint Conference on Neural Networks (IJCNN), pp. 1–7 (2020)
28. Lin, Y., Ting, Y., Huang, Y., et al.: Integration of deep learning for automatic recognition of 2D engineering drawings. Machines **11**(8), 1–20 (2023)
29. Villena Toro, J., Tarkian, M.: Automated and Customized CAD Drawings by Utilizing Machine Learning Algorithms: A Case Study. Volume 3B: 48th Design Automation Conference (DAC), 1-10 (2022)
30. Kirillov, A., Mintun, E., Ravi, N., et al.: Segment Anything, pp. 1–30 (2023)
31. He, K., Zhang, X., Ren, S. et al.: Deep Residual Learning for Image Recognition, pp. 1–12 (2015)
32. Ren, S., Girshick, R., Sun, J.: Faster R-CNN: Towards Real-Time Object Detection with Region Proposal Networks, 1–14 (2015)
33. Ye, J., Hu, A., Xu, H. et al.: mPLUG-DocOwl: Modularized Multimodal Large Language Model for Document Understanding, pp. 1–10 (2023)
34. Laurençon, H., Tronchon, L., Cord, M., et al.: What matters when building vision-language models? pp. 1–26 (2024)
35. Li, M., Lv, T., Chen, J. et al.: TrOCR: Transformer-based Optical Character Recognition with Pre-trained Models, pp. 13094–13102 (2023)

Certainty in Uncertainty: Exploring Probabilistic Approaches in AI

Uncertainty Estimation of Raters' Performance and Ground Truth Through a Bayesian Extension of STAPLE

Davide Cazzorla and Corrado Mencar[✉]

Department of Computer Science, University of Bari Aldo Moro, Bari, Italy
{davide.cazzorla,corrado.mencar}@uniba.it

Abstract. We tackle the problem of merging information originating from several imperfect raters in labelling items, without the availability of a ground truth. This problem can be approached by STAPLE, which is an algorithm for estimating the ground truth and assessing the quality of raters based on Expectation-Maximization. However, the results of STAPLE are precise estimations, which hide the uncertainty on the true values. In this paper, we introduce a fully Bayesian extension of STAPLE, which provides posterior distributions of the ground truth and the raters' performance. Sampling is based on Gibbs method to ensure fast estimation even with large data. Experimental results show that the Bayesian extension uncovers some potential bias of the original STAPLE, and offers a representation of uncertainty to help the decision maker in assessing the reliability of the estimations.

Keywords: STAPLE · Bayesian modeling · ground truth estimation

1 Introduction

In some applications like medical image segmentation, a number of raters (humans or algorithms) are asked to label items to estimate a missing ground truth. For example, raters may segment the image of the slice of a brain to isolate some detected lesions. Usually, by merging all information of the raters, a final label is eventually established. However, raters are usually inexact, thus ground truth can only be inferred from statistical considerations of independent raters [7].

STAPLE (Simultaneous Truth And Performance Level Estimation) is an algorithm for estimating the ground truth and assessing the quality of raters, at the same time [11]. The ground truth is estimated by maximizing the likelihood of observed data provided by raters, through Expectation Maximization (EM). Although STAPLE was originally conceived for image segmentation (thus, taking into account spatial correlation of 3D voxels), its framework is very general and can be applied to a wide variety of learning problems [9].

STAPLE has been extended in several ways to overcome the limitations of the original version. One limitation of the original method stands in the estimation of the ground truth, which depends on a *precise* estimation of the performance of the raters. While this assumption enables a fast estimation of the ground truth, it does not take into account the uncertainty on the performance of the raters. However, a quantification of performance uncertainty is important to understand whether more data (i.e., more raters) are needed for a more certain estimation of the ground truth.

The problem of assessing uncertainty of performance was tackled through the estimation of confidence intervals associated to the raters' performance [5]. However, these confidence intervals are estimated in a post-processing stage, therefore they do not influence the estimation of the ground truth, which is still based on EM. The proposed study is based on Bayesian analysis, with the specific aim of introducing performance uncertainty *within* the estimation process, in order to propagate uncertainty to the estimated ground truth.

Bayesian extensions of STAPLE are available in literature. Specifically, Variational Bayes was used to fuse image segmentations represented in terms of probability maps that are transformed into a Euclidean space, onto which Gaussian distributions are used to perform estimation [3]. Empirical Bayes was used to estimate the prior distribution of the ground truth, then EM was applied to find the optimal parameters of the model [1]. Finally, Bayesian analysis was adopted to show some theoretical features and limitations of STAPLE [10].

Differently from previous work, this paper proposes a fully Bayesian approach to extend STAPLE in order to provide an estimation of uncertainty on the ground truth and the performance of raters. The goal is to enrich the information provided as a result of the fusion process with uncertainty quantification, which could help decision makers in accepting estimations or requiring additional data. Additionally, to cope with the unavoidable increase of computational demand, Gibbs sampling has been adopted for fast estimation of probability distributions.

An experiment with synthetic data shows the advantage of the Bayesian extension of STAPLE over a precise estimation; while an experiment on medical image segmentation shows that precise estimation could be biased, while the Bayesian extension may give more insight on the unknown ground truth.

2 Method

2.1 Preliminaries

A collection of N items, each of unknown class $T_i \in \{0,1\}, i = 1, 2, \ldots, N$, are classified by R raters that assign a binary label $D_{ij} \in \{0,1\}$ for $i = 1, 2, \ldots, N$ and $j = 1, 2, \ldots, R$. As a guiding example, the items correspond to the voxels of an image that is segmented by raters to detect some anatomical regions of interest; each voxel is labelled as '1' if it belongs to a region, '0' if it belongs to the background. Let **D** be the $N \times R$ matrix arranging the classifications of all the raters, and **T** the $N \times 1$ unknown vector representing the ground truth.

It is assumed that each rater has some aleatoric uncertainty in classifying items. This is modeled by introducing raters' *sensibility* p_j and *specificity* q_j, defined as:

$$p_j = \Pr(D_{ij} = 1 | T_i = 1) \tag{1}$$

and
$$q_j = \Pr(D_{ij} = 0 | T_i = 0) \qquad (2)$$

Let $\mathbf{p} = (p_1, p_2, ... p_R)$ and $\mathbf{q} = (q_1, q_2, ... q_R)$ the $R \times 1$ unknown vectors representing sensibility and specificity of all raters, respectively.

Uncertainty on $(\mathbf{p}, \mathbf{q})$ and the ground truth $\mathbf{T}$ is modeled by considering them as random variables. It is assumed that raters are independent. For the sake of simplicity, we will also assume that the ground truth of each item is independent w.r.t. all other items; this assumption can be very strong in some applications, such as image segmentation, and can be loosened by considering a Markov Random Field (MRF) to represent spatial correlation among voxels. The introduction of MRF to the proposed method does not increase the complexity of the model too much, but we preferred to not use it so as to better highlight the differences between STAPLE and its Bayesian extension.

2.2 STAPLE

The goal of STAPLE is to estimate the values of $(\mathbf{p}, \mathbf{q})$ to maximize the complete data log likelihood function (MLE):

$$(\hat{\mathbf{p}}, \hat{\mathbf{q}}) = \arg\max_{\mathbf{p},\mathbf{q}} \ln f(\mathbf{T}, \mathbf{D} | \mathbf{p}, \mathbf{q}) \qquad (3)$$

where f denotes the probability density function of the involved random variables. Since the ground truth $\mathbf{T}$ is unknown, the estimation of $(\hat{\mathbf{p}}, \hat{\mathbf{q}})$ requires an optimization algorithm, which corresponds to Expectation Maximization (EM) in STAPLE.

The maximization of (3) requires the definition of a prior distribution of the ground truth $f(T_i)$. In the case that all items are assumed to be independent, the prior distribution can be estimated by averaging all observed labels over all items and raters:

$$\Pr(T_i = 1) = \frac{1}{NR} \sum_{j=1}^{R} \sum_{i'=1}^{N} D_{i'j} = w \qquad (4)$$

In more realistic scenarios, a MRF is introduced to model the spatial correlation of items. Additional details on the specification of STAPLE can be found in the original work [11].

2.3 Bayesian Extension

We are interested in the estimation of the ground truth $\mathbf{T}$ as well as the performance of the raters $\mathbf{p}$ and $\mathbf{q}$, given the observations $\mathbf{D}$. According to Bayes rule, we have:

$$f(\mathbf{p}, \mathbf{q}, \mathbf{T} | \mathbf{D}) = \frac{f(\mathbf{D}|\mathbf{T},\mathbf{p},\mathbf{q}) f(\mathbf{T},\mathbf{p},\mathbf{q})}{f(\mathbf{D})} \qquad (5)$$

with
$$f(\mathbf{T}, \mathbf{p}, \mathbf{q}) = f(\mathbf{T}) \cdot f(\mathbf{p}) \cdot f(\mathbf{q})$$

because we assume independence of ground truth with respect to expert performance, as well as independence of sensibility and specificity of raters. We also assume independence of raters and independence among items. The latter assumption may be very strong in some applications, e.g., when items represent spatially correlated voxels in medical imaging. Such assumption can be weakened by augmenting the model with a sub-model to represent dependencies, such as a Markov Random Field. We are not adopting this solution in this work for the sake of conciseness.

According to Eqs. (1) and (2), parameters p_j and q_j are probabilities, therefore their prior distributions can be conveniently modeled by Beta distributions:

$$p_j \sim \text{Beta}(\alpha_{p_j}, \beta_{p_j}), \quad q_j \sim \text{Beta}(\alpha_{q_j}, \beta_{q_j}) \tag{6}$$

while the prior of T_i can be either a constant value w as in the original STAPLE, see (4), or can be defined as a Beta distribution:

$$w \sim \text{Beta}(\alpha_w, \beta_w) \tag{7}$$

In such a case, we extend the model (5) to a *hierarchical model* defined as:

$$f(\mathbf{p}, \mathbf{q}, \mathbf{T}, w | \mathbf{D}) = \frac{f(\mathbf{D}|\mathbf{T}, \mathbf{p}, \mathbf{q}, w) f(\mathbf{T}, \mathbf{p}, \mathbf{q}|w) f(w)}{f(\mathbf{D})} \tag{8}$$

and assume pairwise independence between $\mathbf{T}, \mathbf{p}, \mathbf{q}, w$. Notice that the parameter w is the same for all items T_i; more sophisticated models could involve item-wise parameters w_i for $i = 1, 2, \ldots, N$.

The posterior distribution (8) can be estimated numerically via Markov Chain Monte Carlo (MCMC). The structure of the model suggests the adoption of Gibbs sampling for estimating the posterior, which enables the estimation of a joint distribution through its conditionals. In fact, we observe

$$T_i | \mathbf{D}_i, \mathbf{p}, \mathbf{q}, w \sim \text{Bernoulli}(\theta)$$

where:

$$\theta = \Pr(T_i = 1 | \mathbf{D}_i, \mathbf{p}, \mathbf{q}, w)$$
$$= \frac{\prod_j p_j^{D_{ij}} (1-p_j)^{(1-D_{ij})} w}{\prod_j p_j^{D_{ij}} (1-p_j)^{(1-D_{ij})} w + \prod_j (1-q_j)^{D_{ij}} q_j^{1-D_{ij}} (1-w)}$$

We also observe that:

$$f(p_j | \mathbf{T}, \mathbf{q}, \mathbf{D}, w) \propto f(p_j, \mathbf{T}, \mathbf{q}, \mathbf{D})$$
$$\propto p_j^{\alpha_{p_j} - 1 + \sum_i D_{ij} T_i} (1-p_j)^{\beta_{p_j} - 1 + \sum_i (1-D_{ij}) T_i}$$

Therefore,

$$p_j | \mathbf{T}, \mathbf{q}, \mathbf{D}, w \equiv p_j | \mathbf{T}, \mathbf{D} \sim \text{Beta}\left(\alpha_{p_j} + \sum_i D_{ij} T_i, \beta_{p_j} + \sum_i (1-D_{ij}) T_i\right)$$

With similar arguments, it is possible to observe that:

$$q_j|\mathbf{T},\mathbf{p},\mathbf{D},w \equiv q_j|\mathbf{T},\mathbf{D} \sim \text{Beta}\left(\alpha_{q_j} + \sum_i(1-D_{ij})(1-T_i),\ \beta_{q_j} + \sum_i D_{ij}(1-T_i)\right)$$

In the case that w is stochastic, then the conditional probability distribution is

$$f(w|\mathbf{T},\mathbf{p},\mathbf{q},\mathbf{D}) \propto f(\mathbf{T}|w,\mathbf{p},\mathbf{q},\mathbf{D})f(w)$$

$$\propto \prod_i w^{T_i}(1-w)^{1-T_i} \frac{w^{\alpha_w-1}(1-w)^{\beta_w-1}}{B(\alpha_w,\beta_w)}$$

therefore:

$$w|\mathbf{T},\mathbf{p},\mathbf{q},\mathbf{D} \equiv w|\mathbf{T} \sim \text{Beta}\left(\sum_i T_i + \alpha_w,\ \beta_w + \sum_i(1-T_i)\right) \quad (9)$$

2.4 Initialization of Hyper-Parameters

The model parameters p_j, q_j and w require the specification of hyper-parameters $\alpha_{p_j},\beta_{p_j},\alpha_{q_j},\beta_{q_j}$ and α_w,β_w respectively. These parameters can be initialized according to a data-driven (Empirical Bayes) or knowledge-driven approach.

In this work we consider the knowledge-driven approach, which prescribes to inject knowledge through priors when available, or to use the least possible informative priors when knowledge is not available. If there is no knowledge, then $\alpha_{p_j} = \beta_{p_j} = \alpha_{q_j} = \beta_{q_j} = \alpha_w = \beta_w = 1$ is set, so all parameters are uniformly distributed between 0 and 1. In this case, to avoid the label switching problem, where sensitivity and specificity assume values near 0, it is enough to sample p_j^0 and q_j^0 in the Gibbs sampler from a Uniform distribution in the range $[0.5, 1]$.

3 Experimental Results

3.1 Original STAPLE vs. Bayesian Extension

We used synthetic data depicted in Fig. 1. Each image represents a square of 10×10 pixels, which has been segmented into two regions (white and black) according to a rater. All raters agree that the central 2×2 square is a region of interest; while, according to rater 1 the region of interest is 1 pixel wider.

For both STAPLE and Bayesian extension STAPLE, the prior $w = \Pr(T_i = 1)$ is fixed to 0.2. For the Bayesian extension, all hyper-parameters have been initialized so that the priors are the least informative. Posterior distribution has been estimated through Gibbs sampling, with 5,000 iterations (excluding burn-in) and 100 chains, so as to yield a precise estimation of the posterior. (In practice, much fewer iterations and chains are enough.)

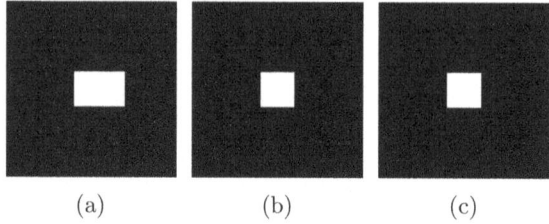

(a) (b) (c)

Fig. 1. Synthetic dataset, made of three 10×10 pixel squares, one per rater. White pixels correspond to $T_i = 1$. (a) Rater 1 thinks that the area of interest is 1 pixel wider than the area detected by rater 2 (b) and 3 (c), which are identical.

Noticeably, STAPLE converges to two different solutions depending on its initialization. In one solution, $\mathbf{p} = [1, 0.66, 0.66]$, which correspond to the case that rater 1 is actually correct, while raters 2 and 3 are missing two out of six pixels, corresponding to the left rectangle of Fig. 1a. In the alternative solution, $\mathbf{p} = [1, 1, 1]$, meaning that all the three raters correctly detected the ground truth made of the 2×2 white square of Fig. 1b/c. (In this solution, the over-segmentation of rater 1 is not affecting their sensitivity.) It is important to notice that STAPLE is providing *either the first or the second solution*; however, both are consistent with data, because ground truth is not available.

Figure 2 reports the posterior distributions of $f(p_j|\mathbf{D})$. Accordingly, the sensitivity parameters are all uncertain due to the small dataset; however, the sensitivity of rater 1 is more specific than the other raters, because their sensitivity could be less than 1 if the ground truth corresponds to the segmentation of rater 1. Thus, the Bayesian extension of STAPLE is capable of capturing the uncertainty related to the ground truth, without collapsing to an arbitrary solution. This granular representation can be used to decide whether additional data (e.g., additional raters) must be included in the data analysis workflow.

3.2 Flat vs. Hierarchical Model

We used the MSSEG 2016 dataset, which contains MRI images from 4 centers, each made using a different machine [6]. Data from five patients are available for each of the centers. For each patient there are different MRI images obtained with different types of imaging sequences, but in this paper, only the images (a.k.a. slices) obtained by FLAIR sequence are used. Also present for each patient are segmentations made by 7 junior neuro-radiologists after they took a course on how to identify multiple sclerosis lesions in MRI images. For this dataset, it is assumed that the neuro-radiologists performed the image segmentation independently, although a dependence between the segmentations of some of them has been hypothesized [8]. In this work, we only considered patient #1 of center #1. Segmentations of slice #150 are depicted in Fig. 3.

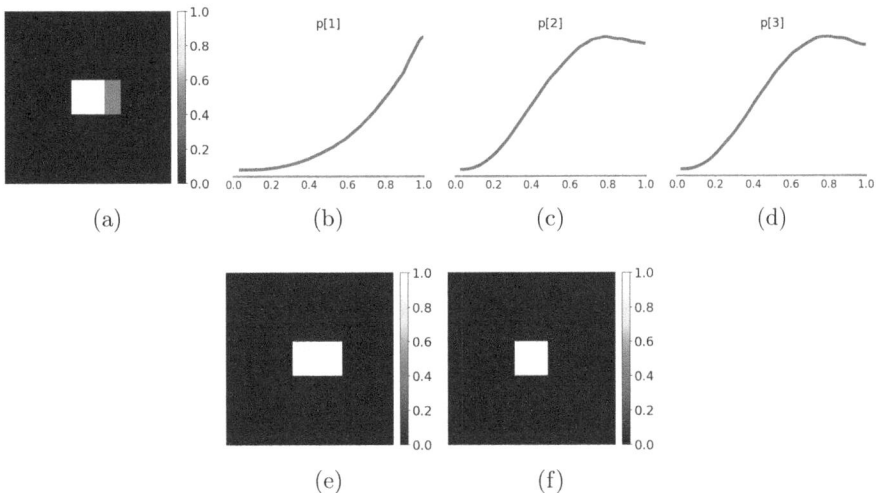

Fig. 2. (a) Ground truth estimated by Bayesian STAPLE. The value of each pixel is given by $\Pr(T_i = 1|\mathbf{D})$ and estimated from MCMC traces. (b–d) Posterior distributions $f(p_j|\mathbf{D})$. (e) Estimated ground truth by STAPLE with initialization $p_1^0 = 0.99, p_2^0 = p_3^0 = 0.7$. (f) Estimated ground truth by STAPLE with initialization $p_1^0 = p_2^0 = p_3^0 = 0.99$. Every pixel is colored according to $\Pr(T_i = 1|\mathbf{D}, \hat{\mathbf{p}}, \hat{\mathbf{q}})$, where $(\hat{\mathbf{p}}, \hat{\mathbf{q}})$ is the MLE estimation of sensitivity and specificity.

In order to perform Bayesian STAPLE without a hierarchical model, it is necessary to define the value of the hyper-parameter w, which is the probability that the ground truth voxels contain the anatomical structure, i.e., $\Pr(T_i = 1)$. The STAPLE paper suggests using the formula (4). With this formula we are assuming that the number of voxels in the ground truth is an average of the number of voxels in the expert segmentations, but this may not be correct and can introduce bias in the estimate. Using the hierarchical model described in (8), this problem is solved because the value of w is updated at each iteration of Gibbs sampling using the ground truth estimate. In fact, the full conditional (9) depends on the number of voxels labeled as one and zero in the ground truth.

For this experiment, the segmentations shown in Fig. 3 will be used. Using the Eq. (4), the hyper-parameter w was initialized to 0.01552. Using the hierarchical model (8) the hyper-parameters of the prior distribution of w were initialized as $\alpha_w = 1, \beta_w = 1$, which is equivalent to having a Uniform distribution in $[0, 1]$. For both models, the hyper-parameters of the prior distributions for sensitivity and specificity were defined as $\alpha_{p_j} = \alpha_{q_j} = 50$ and $\beta_{p_j} = \beta_{q_j} = 1$, $\forall j \in 1, \ldots, 7$. For achieving convergence, 20 chains of 500 iterations were performed. For each chain, the first 100 iterations were discarded as burn-in.

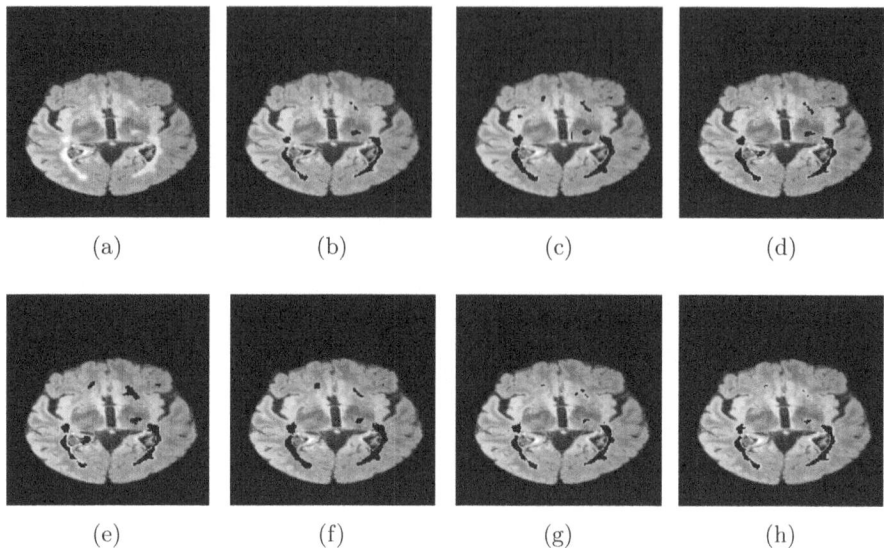

Fig. 3. (a) Slice #150 without segmentations. (b)–(h) Segmentations made by experts: red shade correspond to detection of multiple sclerosis lesions ($T_i = 1$). (Color figure online)

Figure 4 shows the results of Bayesian STAPLE runs with and without the hierarchical model. From the two ground truth estimates, it can be seen that they are slightly different. In particular, with the hierarchical model, at the edges of the lesion this is more likely to be present (the color is light blue instead of blue). In fact, from Fig. 4c and 4d, it is apparent that Bayesian STAPLE is giving higher values to w, though with some uncertainty. More interestingly, the estimate of original STAPLE is significantly outside the range estimated by Bayesian STAPLE, thus highlighting the bias in the estimation; in fact, using Eq. (4), it is assumed that all raters are correct in labelling the voxels, while the hierarchical model takes into account the estimation of the ground truth which, in turn, depends on the estimation of sensitivity and specificity of raters.

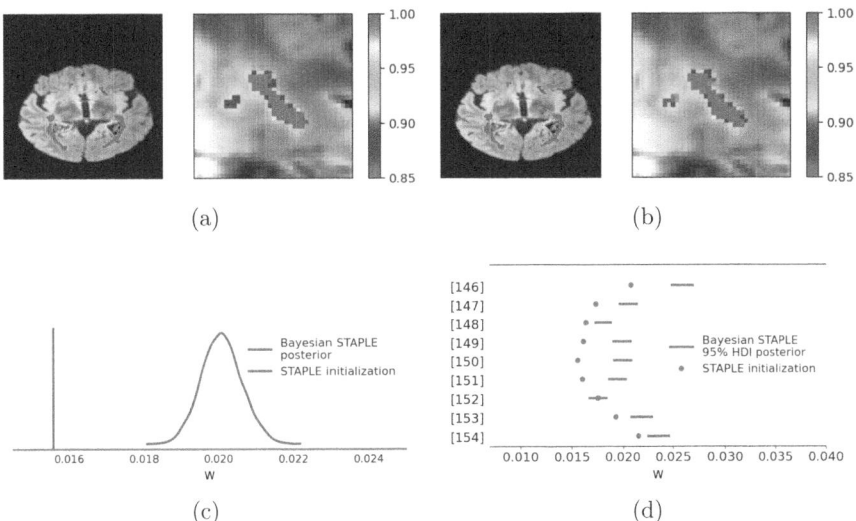

Fig. 4. Estimation of ground truth (a) without hierarchical model, and (b) with hierarchical model. (Entire image at the left, a zoomed-in detail at the right.) Voxel color denotes $\Pr(T_i = 1|\mathbf{D})$. Sub-figs (a) and (b) report voxel colors only when $\Pr(T_i = 1|\mathbf{D}) > 0.01$. Diagram (c) shows the posterior $f(w|\mathbf{D})$ using the hierarchical model. The vertical line represents the estimate of w using the original STAPLE method. Sub-fig (d) shows the same information as (c) but for different slices (y-axis). The blue lines are the 95% Highest Density Intervals for $f(w|\mathbf{D})$. (Color figure online)

4 Conclusion

In this work, a Bayesian extension of STAPLE is proposed, which allows for several advantages and features. Foremost among these is the ability to estimate ground truth using all possible parameter values, weighted by the probability of being the true value. In addition, it has been shown from the experimental results that no information is lost with this approach than with STAPLE. A second advantage is the possibility of not having to set the prior probability for ground truth. Also, the results show that using the hierarchical model for estimating the ground truth turns out to be less biased than initializing the prior probability through the data. A third advantage, is the ability to estimate the uncertainty on the parameters and be able to figure out whether more data are needed.

In this work, it has been assumed that expert performance is equal over the entire item set, but this might be too strong an assumption. In fact, several extensions of STAPLE have been proposed to solve this problem [2–4]. However, this problem has not yet been approached through the Bayesian framework and

Gibbs sampling. In addition, using the Bayesian framework would allow us to calculate the uncertainty for the performance parameters and understand whether more data is needed in some subsets of items, like specific areas of an image.

Moreover, in Bayesian STAPLE with hierarchical extension, it was assumed that each voxel depends on the same hyperparameter w. Again, this assumption may be too strong for the same reasoning that was made for the performance parameters in the previous point. That is, it can be assumed that there is not just one global hyperparameter, but there are several that affect different subsets of the itemset.

Finally, in this work, only some extensions of the original STAPLE article were considered. Future work could consider some of these extensions, which take into account the correlation of items, and the possibility of non-binary ratings. Future work will also consider a quantitative comparison with other Bayesian extensions available in literature.

Acknowledgments. Corrado Mencar is member of the Gruppo Nazionale Calcolo Scientifico-Istituto Nazionale di Alta Matematica (GNCS-INdAM).

References

1. Akhondi-Asl, A., Warfield, S.K.: Estimation of the prior distribution of ground truth in the STAPLE algorithm: an empirical Bayesian approach. In: Ayache, N., Delingette, H., Golland, P., Mori, K. (eds.) MICCAI 2012. LNCS, vol. 7510, pp. 593–600. Springer, Heidelberg (2012). https://doi.org/10.1007/978-3-642-33415-3_73
2. Asman, A.J., Landman, B.A.: Formulating spatially varying performance in the statistical fusion framework. IEEE Trans. Med. Imaging **31**(6), 1326–1336 (2012). https://doi.org/10.1109/TMI.2012.2190992
3. Audelan, B., Hamzaoui, D., Montagne, S., Renard-Penna, R., Delingette, H.: Robust Bayesian fusion of continuous segmentation maps. Med. Image Anal. **78**, 102398 (2022). https://doi.org/10.1016/j.media.2022.102398
4. Commowick, O., Akhondi-Asl, A., Warfield, S.K.: Estimating a reference standard segmentation with spatially varying performance parameters: local MAP STAPLE. IEEE Trans. Med. Imaging **31**(8), 1593–1606 (2012). https://doi.org/10.1109/TMI.2012.2197406
5. Commowick, O., Warfield, S.: Estimation of inferential uncertainty in assessing expert segmentation performance from STAPLE. IEEE Trans. Med. Imaging **29**(3), 771–780 (2010). https://doi.org/10.1109/TMI.2009.2036011
6. Commowick, O., et al.: Multiple sclerosis lesions segmentation from multiple experts: the MICCAI 2016 challenge dataset. Neuroimage **244**, 118589 (2021). https://doi.org/10.1016/j.neuroimage.2021.118589
7. Joskowicz, L., Cohen, D., Caplan, N., Sosna, J.: Inter-observer variability of manual contour delineation of structures in CT. Eur. Radiol. **29**(3), 1391–1399 (2019). https://doi.org/10.1007/s00330-018-5695-5
8. McKinley, R., et al.: Simultaneous lesion and brain segmentation in multiple sclerosis using deep neural networks. Sci. Rep. **11**(1), 1087. https://doi.org/10.1038/s41598-020-79925-4
9. Raykar, V.C., et al.: Learning from crowds. J. Mach. Learn. Res. **11**(4) (2010)

10. Van Leemput, K., Sabuncu, M.R.: A cautionary analysis of STAPLE using direct inference of segmentation truth. In: Golland, P., Hata, N., Barillot, C., Hornegger, J., Howe, R. (eds.) MICCAI 2014. LNCS, vol. 8673, pp. 398–406. Springer, Cham (2014). https://doi.org/10.1007/978-3-319-10404-1_50
11. Warfield, S.K., Zou, K.H., Wells, W.M.: Simultaneous truth and performance level estimation (STAPLE): an algorithm for the validation of image segmentation. IEEE Trans. Med. Imaging **23**(7), 903–921 (2004). https://doi.org/10.1109/TMI.2004.828354

Uncertainty Estimation for Energy Consumption Nowcasting

Danel Rey-Arnal[1](✉)[iD], Ibai Laña[1](✉)[iD], and Pablo G. Bringas[2][iD]

[1] Tecnalia, 48160 Derio, Spain
{danel.rey,ibai.lana}@tecnalia.com
[2] University of Deusto, 48007 Bilbao, Spain

Abstract. In recent years nowcasting systems have been required to perform on more complex scenarios and to a better standard than ever. This work aims at studying the relation between the performance of an array of prediction models improved via data aggregation and the measurement of uncertainty and its potential shift as a consequence of such aggregation. In order to gauge the impact of this approach we propose an experimental framework in which the predictive capabilities of distinct modeling approaches in different aggregation circumstances, is assessed in conjunction with the study of model uncertainty measurements. The results show that the not only the aggregation level, but the modeling choice can have an impact in terms of uncertainty quantification, revealing different sizes of confidence intervals. These measurements represent a novel approach to the model and aggregation level selection process.

Keywords: Nowcasting · Uncertainty · Aggregation

1 Introduction

Community Energy Systems (CESs) have been studied in different forms over the years, from their initial conception as a co-generation, heat and cold storage system to the modern approach mainly focused on the self-sustaining renewable energy usage [17]. The most commonly considered integrating elements in these studies are batteries and renewable generator arrays (solar photovoltaic panels, wind turbines). One of the main reasons batteries get such levels of attention in the literature comes from the fact that in CESs, community battery storage systems offer greater benefits compared to individual ones [4]. Therefore a great deal of effort is currently being spared towards optimisation methods and simulation tools for the planning and operation of microgrids and CESs as written in [22].

The ability to predict both energy generation and consumption becomes imperative in order to successfully manage such CESs and predictive systems based on Machine Learning have been proven a reliable tool to do so. In terms of energy demand prediction techniques, multi-layer perceptron (MLP) schemes have seen success [6], as well as in for energy consumption [1], although many other non-linear models are being used, including decision trees, such as Random Forest (RF) [2]. Amongst the variables governing household energy consumption, climatological features are one of the most important as they allow for a general scheme of the energetic behaviour of a household to be drawn, nevertheless, energy demand and consumption prediction remains a challenge due to the wide variety of additional influencing factors, such as the diversity of residential sector behaviours, physical properties [27], as well as the type of energy used [8]. Furthermore, the unpredictability of climatological features in a short term scale is a challenge on itself as spontaneous changes in the weather conditions may weaken the reliability of energy generation as well as introduce additional consumption demands such as heating requirements. In addition, temporal variations in household electrical consumption are higher compared to other sectors (such as commercial or industrial), which adds another degree of uncertainty in the prediction making its application less reliable in the case of optimisation environments. To study this uncertainty confidence-based prediction serves as a tool to contextualizes the results.

The aim of this article is to present an array of possible models while considering how both, industry standard performance evaluations and uncertainty estimation, are impacted by the proposed feature aggregation.

2 Related Work

Numerous models have been published as a result of the last ten years of intensive study and scholarly examination of the performance enhancement of power systems nowcasters over different horizons. Nowcasting approaches can be categorized as prediction point (PP) or prediction interval (PI) based on the prediction output they produce [24]. Scientists have worked historically to create new models in order to increase the accuracy of PP predictions. But in recent years, some researchers [32] have begun to argue that PP forecasters only give partial information since they don not provide information regarding potential deviations from the predicted value. Accordingly, these investigations claim that the classic PP's boundary prediction ensures reliability by delivering information to Distribution System Operators (DSOs).

According to current authors Pierre Pinson et al., Huaizhi Wang et al. and Reza Tahmasebifar et al., there are three primary categories of PP forecasters: statistical [23], artificial intelligence (AI) [31], and hybrid models [28]. Amongst the statistical models Kalman filtering [29], auto-regressive integrated moving average (ARIMA) [10], and linear regression are included. The main benefits of statistical models stem from their well-developed algorithms and ease of computation. However, since external elements are often disregarded or because it can be challenging to build linearity between the inputs of the models and the target values, the performance of these models might occasionally be diminished.

In order to tackle the non-linearity found on statistical models, AI emerges as a reasonably viable solution, within the broad family of AI models a distinction can be made between Machine Learning (ML) models and Deep Learning (DL) model. While ML models such as RF, Artificial Neural Networks (ANN) and Support Vector Machines (SVM) have been proven as feasible tools modeling the non-linear behaviour of energy generation and demand time series there exists a historical precedent to be concerned about the models over fitting [3]. Similarly, some of the most popular DL models such as Recurrent Neural Networks (RNN) [18] exhibit a wide array of different use cases as models [26] in spite of the fact that RNNs struggle when accounting for the vanishing gradient effect. The development of the Long Short-Term Memory (LSTM) networks, whose main defect is the risk of a exceedingly great loss in accuracy when dealing with a model of high quantities of input variables and time steps [20], address the flaw presented in RNNs.

While the efficacy of hybrid models in generating precise predictions is well-established, it is equally imperative to comprehend the fact that these forecasts are often associated with a degree of uncertainty. This setback is usually addressed with the implementation of confidence intervals. Confidence interval prediction models are being used at an increasing rate in state-of-the-art methods on a wide array of fields such as energy demand or traffic forecasting [19]. These systems produce results defining both, a confidence interval, a distance over which an α amount of outputs get placed in the distribution and the size of said span, often correlating to the amount of information inferable from the results.

The uncertainty that is inherent in the output of the model when tested with new data is generally divided into two main categories: *epistemic uncertainty*, which includes the uncertainties introduced by the modeling process itself, and *aleatory uncertainty*, which pertains to the immutable uncertainties arising from the natural variability in the data [5]. The precise count of unique uncertainty sources and the extent to which they can be separated are subjects of ongoing debate where statisticians often engage on discussion [12]. Despite the ambiguity surrounding the possibility of distinguishing different sources of uncertainty or isolating the uncertainty associated with data from other data attributes, the application of uncertainty estimation methods remains valuable for relevant ML tasks such as model selection, data preprocessing, or active learning. The quantification of the innate uncertainty can be learnt through the statistical study of the variance of each PP or the use of alternative statistical measures related to the output of the model, such as confidence intervals [11].

A variety of techniques and methods for quantifying uncertainty have been developed by researchers, with a primary focus on imposing the models to yield outputs with a certain level of variance. Bayesian methods are particularly noteworthy as they directly generate probability distributions, thereby eliminating the need for separate steps to estimate uncertainty. Once a distribution is formed, it is possible to calculate statistical significance and confidence intervals. Some machine learning models provide the flexibility to alter their default behavior to produce predictions based on percentiles. This includes ensemble methods (which derive the output of each estimator), test dropout (which introduces variations in the output), quantile regression (which predicts different percentiles), and other similar techniques. These model-specific approaches are often referred to as *intrinsic* in the literature [15]. On the other hand, *extrinsic* uncertainty estimation techniques are not linked to a specific machine learning model but enable the measurement of uncertainty after a model has been trained, typically via a calibration process. Among these techniques, Conformal Prediction has been selected as the reference uncertainty estimation method for this work. The key component of Conformal Prediction is the sturdy calibration process, that secures strength of the yielded intervals, as well as the proven versatility across numerous algorithms. As per Vovk [30] Conformal Predictors exist as a part of confidence predictors.

These advancements in the use of confidence intervals along with the development of LSTM models and LSTM-hybrid models are contributing to more robust and reliable predictions in various fields, including energy demand nowcasting and stock market prediction [9].

3 Materials and Methods

The experiments for this article have been done making use of the *UMass Smart* Data Set for Sustainability*[1]. This open dataset provides aggregated electrical consumption timestamped data of 114 apartments in California. Consumption readings are provided each minute, so a complete day results in 1440 measurements. This fine granularity is convenient in terms of data quantity, but it comes with a set of disadvantages. A 60 s per entry indexation of the information results on a quite variable dataset in which patterns are almost indistinguishable, and thus weakening the prediction. Moreover, there is a strong case against a 1 min time horizon, if the *actionability* of the prediction is considered: if a good forecast was possible in this aggregation interval, an operative action would be hard to support on a proper response time.

Henceforth, the data have been preprocessed for the experiments. In the first place, the raw data were cleaned, removing data points that were not available for all apartments. Besides a missing data study and imputation was performed. The second step involves temporal and spatial aggregation of the raw data. Temporal aggregation involves changing the aggregation level of the 1 min signal into different aggregation steps (5,10,15,30 and 60 min). As electrical consumption is additive, this aggregation has been performed by means of summing the 1 min values recorded during the new aggregation intervals. This allows for the grounds of our study, that is aimed at discovering the sweet spot of aggregation that allows models to provide good forecasts without loosing information about the underlying signals.

Besides the temporal aggregation, a spatial aggregation is performed over apartments. This aggregation operates in the same direction than temporal one: the more apartments together, the smoother the signal, as the load profile of lower aggregation levels is less prone to significant events which leads to extreme spiking from a flat line along the day, as the different appliances get started or

---

[1] https://traces.cs.umass.edu/index.php/Smart/Smart.

plugged, depending on the life schedule of the household's inhabitants. It also has the same drawback, loosing detailed information of apartments. Another consequence, as the aggregation is performed by addition, is the change in the dynamic range of the signal, which needs to be considered in the result evaluation. The spatial aggregation is performed via a combine and perturb method in which for an apartment aggregation level (e.g. 5 apartments), different random combinations of that number of apartments are performed in order to guarantee statistical significance of the results, since in a real situation scenario apartments with random behavior are aggregated on a particular circuit breaker panel. Thus 6 levels of apartment aggregation (3, 5, 10, 15, 20 and 25 apartments) are considered for the study, aggregation that will operate jointly with the temporal aggregation described above.

Once the datasets are ready, they are split into three subsets, 72% of the data (80% of the 90% of the original training split) is used in training while 18% (20% of the 90% training data) is destined for the calibration of the confidence intervals and the remaining 10% is reserved for testing purposes.

3.1 Experimental Design

The proposed experimental framework includes performing predictions on all of the above described datasets and follows the scheme depicted in Fig. 1.

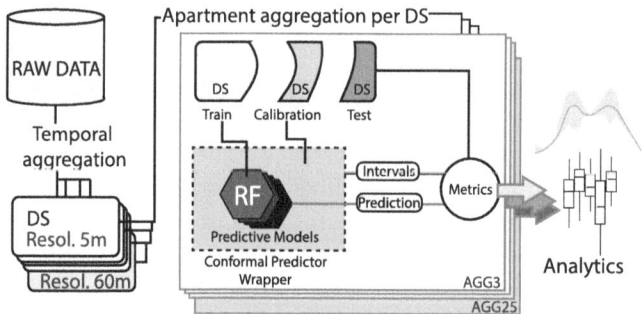

Fig. 1. Graphical representation of the experimental sequence

Once the datasets are defined and the train, calibration and test splits are obtained, the experiments involve predicting the consumption in the different spatio-temporal aggregation combinations with different modeling techniques. These techniques are wrapped on a Conformal Prediction framework that will be fitted to obtain calibrated confidence intervals. The selected models are Random Forest (RF[1] [7], Extra Tree Regressor (ETR) [14], Gradient Boost (GB) [21] and a Multilayer Perceptron (MLP) [25]. The initial state and parametrization for these algorithms has been selected via grid search. While RF, ETR and GB are all tree-based models, as such they are less susceptible to outliers than other models, a greatly sought characteristic when consumption spikes are taken into account, however as these models handle feature co-linearity implicitly pruning may be needed if features have a high correlation between each other. This problem is addressed differently in RF and ETR as RF internally bags the data while ETR uses the whole dataset to fit a large amount of unpruned trees. Internally GB builds an additive model which means that by aggregating weak learners it builds a strong one. In each stage a regression tree is fit on the negative gradient of the given loss function [13]. All of these features can have an impact in the uncertainty of their results, reason for which these techniques are compared. On the other hand, MLP trains iteratively since at each time step the partial derivatives of the loss function with respect to the model parameters are computed to update the parameters and can prevent overfitting while being able to handle non-linear problems quite successfully [16]. As seen in Fig. 1, the train and calibration datasets are fed to the Conformal Predictor wrapper and by the end of the process the expected results include a series of metrics such as Interval Coverage Percentage (ICP) and Mean Interval Length (MIL) as well as the performance. These metrics are used to evaluate the performance of the model and to determine if it meets the desired specifications.

To understand the performance of the diverse models we have selected the coefficient of determination, R^2. While nowcasting strategies are usually analyzed through the lens of *Root Mean Squared Error* (RMSE), due to the fact that RMSE will not lead to conclusive evidence as aggregating apartments will greatly alter the dynamic range of the datasets, making them unable to be compared between each other, while R^2 operates on a static range $(-\infty, 1]$, as such it fits better as a performance metric for this work. The mathematical formulation for the coefficient of determination is defined as follows:

$$R^2 = 1 - \frac{\sum (y_i - \hat{y}_i)^2}{\sum (y_i - \bar{y}_i)^2} \tag{1}$$

On Eq. 1 $\hat{y}_i$ represents the expected values of y, while $\bar{y}_i$ is the mean values of y.

On the other hand, the uncertainty quantification process yields its own metrics. *Mean Interval Length* is a metric that shows for a given significance level α the average length of the interval needed to cover the number of values indicated by said α, that has been fixed to 0.9 for our study. Thus, a highly restricted interval length provided in combination with lofty confidence will yield substantial information as a result, conversely a lax interval length will derive on a wide array of potential values causing the obtained data to be of less significance even on a similar level of confidence. Mathematically, MIL can be expressed as:

$$\text{MIL} = \frac{1}{T} \sum_{t=1}^{T} (u_t - l_t). \tag{2}$$

where u_t and l_t represent the upper and lower boundaries of the confidence interval estimated by the technique at hand, and T is the number of instances in the test dataset.

Interval Coverage Percentage is a metric that measures the percentage of items in the test set that are bound to the defined interval. This metric gives an indication of how closely the interval matches the designated significance level (1-α)). The mathematical definition of this metric is:

$$\text{ICP} = \frac{1}{T} \sum_{t=1}^{T} I(l_t \leq y_t \leq u_t). \qquad (3)$$

where matching Eq. 2 u_t and l_t represent the upper and lower boundaries of the confidence interval and I means that only the elements y_t comprehended between the intervals will be used on the summation.

These two metrics give additional context of how the models perform in isolation, allowing us to evaluate the growth of the uncertainty levels as the prediction performance increases. This makes it possible to compare models in a more nuanced way by identifying those that consistently produce narrower intervals (lower MIL), which denotes reduced uncertainty as the estimated possible values for the PI are nearer to the real PP.

4 Experiments and Results

The experiments have been conducted according to the experimental design Fig. 1 presented in the previous section and the overall results are presented in Table 1. An analysis of the performance results of each model and dataset reveals that the difference between models is negligible for most combinations of aggregation and resolutions with MLP as a slightly top-performer on aggregation 25 temporal resolution 60 which sees above average results on a small margin. When comparing the different aggregation levels we can see a clear improvement in performance as the aggregation level grows which is the central notion around this work. This conception can be challenged when comparing aggregation levels 3 and 5 that report very similar results with slight improvements, however we attribute the modest growth to the difference of apartments between aggregation levels as there is only a two apartment discrepancy among these levels while the next levels increase the amount of apartments in groups of five. On the other hand, while the temporal resolution growth should be reflected on the performance, temporal horizons 10 and 15 report lower values (worse results) than horizon 5, our conjecture is than horizons 10 and 15 lose incidental values such as the turning on moment of an electrical appliance, however by the temporal resolution 30 point this occasions get leveled out and thus the performance increases.

As postulated before, the MIL value is increased along with the aggregation levels and the temporal resolution as uncertainty is supposed to increase with the rise of the dynamic range of the signal. Nonetheless, this interval operates

Table 1. Experiment results. Columns show the algorithmic approach and error metrics while rows show the aggregation level and resolution of data

Agg	Resol	RF ICP / MIL / R^2	ETR ICP / MIL / R^2	GBoost ICP / MIL / R^2	MLP ICP / MIL / R^2
3	5	0.91 / 79.99 / 0.43	0.84 / 31.27 / 0.42	0.89 / 63.85 / 0.44	0.90 / 70.92 / 0.42
	10	0.91 / 151.01 / 0.28	0.84 / 63.10 / 0.28	0.89 / 131.22 / 0.30	0.89 / 139.52 / 0.27
	15	0.90 / 196.70 / 0.32	0.85 / 85.53 / 0.31	0.89 / 176.36 / 0.33	0.90 / 192.53 / 0.28
	30	0.90 / 330.92 / 0.35	0.87 / 146.51 / 0.34	0.90 / 301.95 / 0.38	0.90 / 319.82 / 0.33
	60	0.90 / 598.67 / 0.31	0.86 / 271.57 / 0.30	0.89 / 562.60 / 0.34	0.88 / 591.46 / 0.29
5	5	0.91 / 91.85 / 0.42	0.84 / 37.83 / 0.42	0.89 / 79.23 / 0.45	0.90 / 85.07 / 0.43
	10	0.90 / 175.66 / 0.30	0.84 / 75.26 / 0.30	0.90 / 161.27 / 0.32	0.90 / 170.56 / 0.29
	15	0.89 / 228.06 / 0.34	0.85 / 101.70 / 0.34	0.90 / 217.15 / 0.36	0.89 / 226.20 / 0.33
	30	0.90 / 380.53 / 0.45	0.87 / 167.38 / 0.45	0.89 / 337.01 / 0.48	0.90 / 361.13 / 0.44
	60	0.90 / 705.69 / 0.50	0.87 / 293.90 / 0.49	0.90 / 678.50 / 0.52	0.91 / 669.88 / 0.49
10	5	0.90 / 133.42 / 0.50	0.82 / 55.80 / 0.49	0.89 / 123.30 / 0.52	0.89 / 122.32 / 0.52
	10	0.90 / 263.74 / 0.40	0.83 / 113.81 / 0.39	0.90 / 248.19 / 0.42	0.90 / 248.37 / 0.42
	15	0.90 / 354.81 / 0.46	0.85 / 153.54 / 0.46	0.90 / 338.06 / 0.48	0.9 / 336.22 / 0.48
	30	0.91 / 590.95 / 0.57	0.86 / 257.63 / 0.56	0.89 / 542.08 / 0.58	0.91 / 565.75 / 0.59
	60	0.90 / 1096.72 / 0.58	0.87 / 476.83 / 0.58	0.90 / 1057.44 / 0.61	0.90 / 1062.01 / 0.60
15	5	0.90 / 160.75 / 0.56	0.83 / 67.43 / 0.56	0.90 / 153.15 / 0.58	0.90 / 151.67 / 0.59
	10	0.90 / 323.56 / 0.48	0.83 / 137.63 / 0.48	0.90 / 312.01 / 0.50	0.90 / 312.71 / 0.50
	15	0.90 / 437.47 / 0.52	0.84 / 190.82 / 0.52	0.90 / 427.44 / 0.54	0.90 / 431.70 / 0.54
	30	0.91 / 748.71 / 0.62	0.87 / 331.75 / 0.62	0.90 / 712.81 / 0.64	0.90 / 725.29 / 0.64
	60	0.91 / 1507.22 / 0.60	0.86 / 635.28 / 0.61	0.91 / 1439.43 / 0.64	0.90 / 1422.27 / 0.63
20	5	0.91 / 200.05 / 0.62	0.83 / 82.21 / 0.61	0.90 / 189.93 / 0.63	0.90 / 190.46 / 0.63
	10	0.91 / 415.21 / 0.54	0.84 / 172.21 / 0.53	0.90 / 394.21 / 0.55	0.90 / 396.61 / 0.55
	15	0.91 / 568.19 / 0.59	0.85 / 237.64 / 0.59	0.91 / 551.07 / 0.61	0.91 / 555.65 / 0.60
	30	0.90 / 906.82 / 0.67	0.86 / 407.02 / 0.67	0.90 / 868.52 / 0.69	0.91 / 897.88 / 0.68
	60	0.91 / 1800.94 / 0.68	0.87 / 778.88 / 0.68	0.91 / 1763.01 / 0.70	0.92 / 1856.69 / 0.69
25	5	0.90 / 219.67 / 0.65	0.82 / 93.41 / 0.65	0.90 / 211.25 / 0.67	0.90 / 210.28 / 0.67
	10	0.90 / 462.95 / 0.59	0.83 / 191.20 / 0.59	0.90 / 447.03 / 0.61	0.91 / 454.87 / 0.61
	15	0.91 / 626.22 / 0.64	0.85 / 268.13 / 0.64	0.91 / 607.31 / 0.66	0.91 / 614.92 / 0.66
	30	0.91 / 1055.62 / 0.72	0.87 / 464.09 / 0.71	0.90 / 1000.19 / 0.74	0.90 / 1006.36 / 0.73
	60	0.90 / 2093.32 / 0.71	0.87 / 913.10 / 0.71	0.91 / 2131.80 / 0.74	0.92 / 2155.14 / 0.78

differently under the ETR experiments, reporting values approximately half the MIL of the others models for all scenarios. This fact, in conjunction with the reported ICP being lower but still close to the 0.9 reference, leads us to believe that while the assigned interval for ETR has a lower amount the predicted values inside the confidence interval when taking the difference in MIL into consideration it may be the most reliable model of the presented four.

A lower MIL is a better uncertainty metric (less uncertainty) provided the ICP is conserved. In this particular case, however, the ICP lowers slightly as a

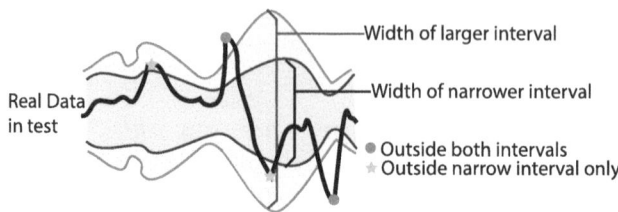

Fig. 2. Comparison on how test samples fall within two different size intervals

consequence of a much narrower interval that leaves more elements out of the boundaries when compared to the rest of the models, that yield a ICP adjacent to the expected 0.90 score. This percentage is reached at the expense of a broader interval as the case for MLP Regressor on aggregation level 25 temporal horizon 60 depicts particularly well, where the ICP raises even higher than the expected 0.90 as the repercussion of gathering more samples than expected constituting a less precise uncertainty measurement. As represented in Fig. 1 the intervals are defined by the training data, which means the real test data can still be found outside the intervals, particularly a percentage of 1-α, provided that the calibration is correct; this means that for a value of $\alpha = 0.9$, we should expect around 10% of samples out of the intervals. An ICP lower than 0.9 implies that the interval is leaving outside some real samples that should be inside. In our results most of the out of interval data is out of both, the wider and the narrower intervals, as the ICP score only deviates slightly between them, this cases are represented by the *circles* in the Fig. 2, while the points represented by the *stars* are the ones that are outside only the narrower interval. In our experiments, for any model, most points would be *circles*, as the differences in ICP suggest that the *star* are just a few samples that lead to that slight ICP difference. The fundamental takeaway is that for this particular experiment as the samples inside the larger interval but outside the narrower interval are a minority, the interval calibrated for the ETR model delivers the greatest amount of information. This would imply that, for a model selection, we could focus on performance, and select neural network based models, or we could focus on having less uncertainty, and thus choose the ETR approach.

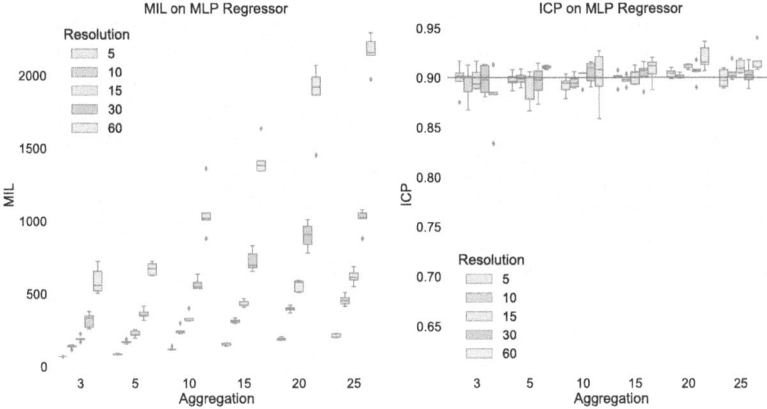

Fig. 3. Mean Interval length and Interval Coverage Percentage while using a Multi-layer perceptron. Boxplots represent the distribution of MIL and ICP of the 5 randomized experiments performed over the aggregation and resolution levels.

Our results show that data aggregation improves the performance of the prediction models as Table 1 illustrates. The boxplots show the distribution of the MIL measured over the 5 experiments done in each spatial (X-axis) and temporal (boxplot colour) aggregation for a single method, MLP. In this Figure we can detect how both, spatial and temporal aggregation add uncertainty progressively, as is expected, as the dynamic range grows in consequence of the aggregation of series being done via addition the interval increases. This analysis will allow us to chose the best combination of spatial and temporal aggregations for a given expected uncertainty value. Thus if the chosen resolution or temporal aggregation is 30 min, there is a noticeable jump between the spatial aggregations of 10 and 20 apartments, hence a spatial aggregation level could be prescribed. Analogously, the second graph on Fig. 3 shows the ICP results for a given method. In this Figure the variability of the measured ICP for each aggregation combination can be discerned. It should be noted that the ICP must stay as adjacent as possible to 0.9, the value selected for the experiments, thus the combinations with the biggest dispersion, such as spatial aggregation 10, are worse results. In addition the combinations that move away from the line such as the spatial aggregations for 20 and 25 apartments, that are way over the selected statistical significance value, implying larger than expected confidence intervals. This analysis makes it possible to assess what are the aggregation levels that produce the most reliable and stable intervals, supporting the prescription of 15 apartment aggregations with a temporal resolution of 10 or 15 min.

5 Concluding Remarks and Future Research

As demonstrated in the results included in Table 1 a higher level of aggregation leads to a better performing model prediction wise, however this performance should be analyzed under the context of the achieved confidence metrics. As both the temporal and spatial aggregation levels increase, there is a higher R^2 score, albeit associated to a bigger uncertainty value, registered in MIL. While at face value it would seem joining a better performing model with a high degree of uncertainty as the worsening of it, it is this additional context that allows us to study the informational efficiency. The informational efficiency is the study of the value brought by data. Moreover, the addition of Conformal Prediction and uncertainty metrics lets us both, study the informational efficiency of our experiment and prescribe the aggregation level recommended for each case. The results reveal additional information from our data, notably, an inflexion point between a reasonable model performance and an adequate uncertainty amount. For our work this point lies between the aggregation levels 15 and 20, particularly on the 30 min temporal resolution as this time frame does not present the problems found on the temporal resolutions 10 and 15 min while maintaining a reasonable operability time span. For different sets of data the distinct combination may vary, however using this technique of uncertainty estimation shows great potential for subsequent studies.

In terms of future research this experiments branches into two possible paths forwards, firstly, an augmented version of this work, in which more prediction models, specifically Deep Learning models such as LSTMs which has great capabilities on capturing non-linearity. However, the calculation of calibrated confidence intervals can be more challenging on deep learning models over the shallow learning models presented in this work. Different aggregation levels, ideally on a neighbourhood or city that already has the apartments aggregated in order to extrapolate our findings to real data may be used. On the other hand we propose an study on load disaggregation where the improved upon predictions can be inferred back to particular apartments.

Acknowledgments.. The authors would like to thank the Basque Government for its funding support through the BIKAINTEK PhD support program (grant no. 014-B2/2022).

Disclosure of Interests. The authors have no competing interests to declare that are relevant to the content of this article.

References

1. Afzal, S., Ziapour, B.M., Shokri, A., Shakibi, H., Sobhani, B.: Building energy consumption prediction using multilayer perceptron neural network-assisted models; comparison of different optimization algorithms. Energy **282**, 128446 (2023)
2. Ahmad, M.W., Mourshed, M., Rezgui, Y.: Trees vs neurons: comparison between random forest and ANN for high-resolution prediction of building energy consumption. Energy Build. **147**, 77–89 (2017)
3. Al Mamun, A., Sohel, M., Mohammad, N., Sunny, M.S.H., Dipta, D.R., Hossain, E.: A comprehensive review of the load forecasting techniques using single and hybrid predictive models. IEEE Access **8**, 134911–134939 (2020)

4. Barbour, E., Parra, D., Awwad, Z., González, M.C.: Community energy storage: a smart choice for the smart grid? Appl. Energy **212**, 489–497 (2018)
5. Beven, K.: Facets of uncertainty: epistemic uncertainty, non-stationarity, likelihood, hypothesis testing, and communication. Hydrol. Sci. J. **61**(9), 1652–1665 (2016)
6. Boersma, K.: Using influencing factors and multilayer perceptrons for energy demand prediction. B.S. thesis, University of Twente (2019)
7. Breiman, L.: Random forests. Mach. Learn. **45**, 5–32 (2001)
8. Chen, M., Ban-Weiss, G.A., Sanders, K.T.: The role of household level electricity data in improving estimates of the impacts of climate on building electricity use. Energy Build. **180**, 146–158 (2018)
9. De, S., Dey, A.K., Gouda, D.K.: Construction of confidence interval for a univariate stock price signal predicted through long short term memory network. Ann. Data Sci. 1–14 (2020)
10. De Felice, M., Alessandri, A., Catalano, F.: Seasonal climate forecasts for medium-term electricity demand forecasting. Appl. Energy **137**, 435–444 (2015)
11. De Jong, G., Daly, A., Pieters, M., Miller, S., Plasmeijer, R., Hofman, F.: Uncertainty in traffic forecasts: literature review and new results for the netherlands. Transportation **34**(4), 375–395 (2007)
12. Der Kiureghian, A., Ditlevsen, O.: Aleatory or epistemic? Does it matter? Struct. Saf. **31**(2), 105–112 (2009)
13. Friedman, J.H.: Greedy function approximation: a gradient boosting machine. Ann. Stat. 1189–1232 (2001)
14. Geurts, P., Ernst, D., Wehenkel, L.: Extremely randomized trees. Mach. Learn. **63**, 3–42 (2006)
15. Ghosh, S., et al.: Uncertainty quantification 360: a holistic toolkit for quantifying and communicating the uncertainty of AI (2021)
16. Kingma, D.P., Ba, J.: Adam: a method for stochastic optimization. arXiv preprint arXiv:1412.6980 (2014)
17. Koirala, B.P., Koliou, E., Friege, J., Hakvoort, R.A., Herder, P.M.: Energetic communities for community energy: a review of key issues and trends shaping integrated community energy systems. Renew. Sustain. Energy Rev. **56**, 722–744 (2016)
18. Kong, W., Dong, Z.Y., Jia, Y., Hill, D.J., Xu, Y., Zhang, Y.: Short-term residential load forecasting based on LSTM recurrent neural network. IEEE Trans. Smart Grid **10**(1), 841–851 (2017)
19. Laña, I., Del Ser, J., et al.: Measuring the confidence of traffic forecasting models: Techniques, experimental comparison and guidelines towards their actionability. arXiv preprint arXiv:2210.16049 (2022)
20. Lv, P., Liu, S., Yu, W., Zheng, S., Lv, J.: EGA-STLF: a hybrid short-term load forecasting model. IEEE Access **8**, 31742–31752 (2020)
21. Mason, L., Baxter, J., Bartlett, P., Frean, M.: Boosting algorithms as gradient descent. In: Advances in Neural Information Processing Systems, vol. 12 (1999)
22. Mendes, G., Ioakimidis, C., Ferrão, P.: On the planning and analysis of integrated community energy systems: a review and survey of available tools. Renew. Sustain. Energy Rev. **15**(9), 4836–4854 (2011)
23. Pinson, P., Madsen, H., Nielsen, H.A., Papaefthymiou, G., Klöckl, B.: From probabilistic forecasts to statistical scenarios of short-term wind power production. Wind Energy Int. J. Progress Appl. Wind Power Convers. Technol. **12**(1), 51–62 (2009)

24. Rodríguez, F., Galarza, A., Vasquez, J.C., Guerrero, J.M.: Using deep learning and meteorological parameters to forecast the photovoltaic generators intra-hour output power interval for smart grid control. Energy **239**, 122116 (2022)
25. Rosenblatt, F.: The perceptron: a probabilistic model for information storage and organization in the brain. Psychol. Rev. **65**(6), 386 (1958)
26. Shi, H., Xu, M., Li, R.: Deep learning for household load forecasting-a novel pooling deep RNN. IEEE Trans. Smart Grid **9**(5), 5271–5280 (2017)
27. Sun, M., Konstantelos, I., Strbac, G.: Analysis of diversified residential demand in london using smart meter and demographic data. In: 2016 IEEE Power and Energy Society General Meeting (PESGM), pp. 1–5. IEEE (2016)
28. Tahmasebifar, R., Moghaddam, M.P., Sheikh-El-Eslami, M.K., Kheirollahi, R.: A new hybrid model for point and probabilistic forecasting of wind power. Energy **211**, 119016 (2020)
29. Takeda, H., Tamura, Y., Sato, S.: Using the ensemble kalman filter for electricity load forecasting and analysis. Energy **104**, 184–198 (2016)
30. Vovk, V., Gammerman, A., Shafer, G.: Algorithmic Learning in a Random World. Springer, New York (2005). https://doi.org/10.1007/b106715
31. Wang, H., Liu, Y., Zhou, B., Li, C., Cao, G., Voropai, N., Barakhtenko, E.: Taxonomy research of artificial intelligence for deterministic solar power forecasting. Energy Convers. Manage. **214**, 112909 (2020)
32. Yang, D., Guo, J.E., Sun, S., Han, J., Wang, S.: An interval decomposition-ensemble approach with data-characteristic-driven reconstruction for short-term load forecasting. Appl. Energy **306**, 117992 (2022)

Knowledge Guided Clustering Medieval Polychromy

Florian Sobieczky[1]($\boxtimes$) and Elisabeth Sobieczky[2]

[1] SCCH - Software Competence Center Hagenberg, Softwarepark 32a,
4232 Hagenberg im Mühlkreis, Austria
`florian.sobieczky@scch.at`
[2] IMAREAL - Institut für Realienkunde des Mittelalters und der frühen Neuzeit,
Körnermarkt 13, 3500 Krems an der Donau, Austria
`elisabeth.sobieczky@plus.ac.at`

Abstract. Expressing the membership of certain events in the historic past as characteristic for a specific cultural era sometimes requires the identification of their accumulated occurrence during a specific period. Many clustering techniques typically applied to the identification of such eras use parameters which cannot be estimated by non-data-scientific users of event data-bases. In the art-historical context of artisan schools applying skills and techniques characteristic of a specific state of cultural achievement, it is necessary to agree on the methodology of such applications defining what is meant by accumulation within a certain epoch. By using a semi-parametric clustering method involving histograms and a parameter readily accessible to specialized art historians, we show that a knowledge guided data scientific approach can surpass the dilemma of parameter-dependence and identify periods in time in which certain artistic techniques have been popular.

Keywords: Non-parametric Clustering · Polychromy of Medieval Wood Sculptures · Knowledge Guided Data Science

1 Introduction

1.1 Clustering of Event Data

Using thresholds for probability density estimates of univariate point-events on the time axis to achieve identification of epochs is an eminent research field [4,6]. Clustering techniques are currently used in cultural heritage research [5]. The discipline of defining the clustering in a reproducible way typically requires a specialized community, e.g., in psychiatry [8], nutrition [10], sustainability [11], and natural habitat conservation [9]. Often, non-parametric, 'self-tuning' methods [7]) are required.

In the present approach, we show how expert knowledge in the art-historical context of the polychromy of wooden medieval sculptures can be used to identify epochs in which the use of certain artistic techniques occurs more than in other times. Namely, an initially semi-parametric clustering technique using histograms is made parametric by involving the expert opinion regarding the quality of certain artisan practices being employed in an accumulated fashion. This identification allows confirming or rejecting hypotheses about what can be considered common artistic practice in certain eras.

1.2 Histograms and Bias-Variance Tradeoff

Determining the bin size of histograms by the minimization of the unbiased Mean Integrated Squared Error (MISE) score is a technique from non-parametric statistics [12, 15]. Optimal bin-sizes of histograms is a prominent research topic in non-parametric statistics [17], and have been proposed in the context of density estimation by Sturges [14], Scott [13], and Freedman and Diaconis [18]. We suggest a modification related to the observed data not consisting of single numbers but intervals representing the possible ranges of the observations.

1.3 Art Historical Context

In the European Early and High Middle Ages, wood sculptures used in ecclesiastic spaces were vividly painted with colors and adorned with precious metals. Sophisticated art techniques achieved different visual and aesthetic effects. Of particular interest in the context of the polychromy of high medieval wooden sculptures is the use of translucent layers of color, which can be determined both on metal leaf (gold leaf, silver leaf, tin foil) and on opaque paint layers through material analysis. The findings were compiled in a research project [3].

While the earlier and contemporary art technological sources (Lucca Manuscript, 8th century; Theophilus, early 12th century, "Heraclius", Book III, late 12th century) describe the application of translucent layers of color to leaf metal, they do not mention a corresponding application to opaque paint layers. On leaf metal, these coatings have the function of producing particularly bright colors ("pictura translucida"). The yellow coating ("lucide") plays a special role, as it serves to intensify the gold tone on gold leaf, but imitates gold on tin foil of silver leaf. On opaque layers of paint, translucent colors not only contribute to the intensification of the color tone and the depth of light, but also to a variety of color nuances in the sense of "varietas colorum" being important as a medieval category of beauty [1, 2].

Against this background, a quantitative evaluation of the compiled data is carried out taking into account the following questions:

1. Can an accumulation of the respective techniques - translucent coatings a) on leaf metal or b) on layers of paint - be identified at certain times?
2. Since only the technique of applying translucent coatings to leaf metals is mentioned in the medieval sources, the question arises as to whether their application to opaque paint layers was practiced at the same time or, for example, only at a later point in time?
3. Can certain periods be identified as typical in the use of translucent coatings by restricting to specific opaque colors?

2 Method

2.1 Histograms and Cross-Validation Score

Our method uses the L^2-optimized histogram theory for univariate events-in-time data $\mathcal{D} = \{X_j\}_{j=1}^n$ [13,18] which minimizes the sum of bias and variance error together with iterated thresholding on the population of the resulting clusters. We think of $\mathcal{D}$ as an i.i.d. sample originating from a distribution with density f supported on the time range $950 - 1300$ CE. Furthermore, the sample points in time are observed with an additional uncertainty: Each sample point in time $X_j \in \mathcal{D}$ is thought of as the center of an interval $[s_j, t_j]$, on which the true occurrence of the sculpture's painting is uniformly distributed as a more precise art-historic date-specification isn't available.

The underlying probability distribution is therefore a mixture of the respective uniform distributions. Classical approaches such as Sturges' rule [14] allow consistent estimators of the bin-width) - but it is known to be important to define optimality in the sense of small MSE for the considered *finite* sample.

$$\widehat{f}(x) = \sum_{j=1}^{m} \frac{w_j}{h} \mathbf{1}_{B_j}(x). \tag{1}$$

The values $\widetilde{N}_j := n \cdot w_k$ are generalization of the simple bin-occupation numbers: They are not just the cardinalities $\#\mathcal{D}_i$ with $\mathcal{D}_i = \mathcal{D} \cap B_i$ and the half-open interval $B_i = [(i-1)h, ih)$ (called N_j in [15], and ν_j in [13] and [12]). Instead of always counting '1' for each point in time falling into bin B_i [16], we count $\frac{1}{t_j - s_j} \int_{s_j}^{t_j} \mathbf{1}_{B_i}(x) dx$ for each data-point X_j in $\mathcal{D}$ which lies in $[s_j, t_j]$, uniformly distributed. Note that this amounts to '1', if $[s_j, t_j] \subset B_i$, but it takes into account the other cases characterized by the bin-size being smaller than the interval on which the sample point is uniformly drawn from. It is doing so by lending w_j the corresponding i'th contribution in the form of the ratio $h/\mu([s_i, t_i])$, i.e. with μ the Lebesgue measure on $\mathbb{R}$ (i.c., the length of the respective interval),

$$w_i = \frac{1}{n} \sum_{j=1}^{n} \frac{\mu(B_i \cap [s_j, t_j])}{\mu([s_j, t_j])}. \tag{2}$$

The effect of this extension of the simple sample-type is that the 'building blocks' of the histogram [13] are thus not of the same height $1/h$. Instead, they depend on the ratio of the length of the support of the uniform distribution of the element of $\mathcal{D}$ to the considered bin size. The risk is the expected loss, $R(m) = \mathbb{E}[L(f, \widehat{f})]$ where as the loss function, the integrated mean square error is used: $L(f, \widehat{f}) = \int (f(x) - \widehat{f}(x))^2 dx$. It is not known exactly, but estimated from the observed sample by the empirical measure. Therefore, the minimum m^* of this empirical risk $\widehat{\mathbb{E}}[L(f, \widehat{f})]$ is defined as the optimal number of bins and determines the (equal) bin size h by the bounded input range T divided by m^*.

Rudemo [15] derived a formula for the minimiser m^* for a sample of point-observations based on the risk function being evaluated by individually integrating each of the three terms resulting from taking the square in $L(f, \widehat{f}_n)$ and by the cross-term involving f and $\widehat{f}_n$ being written as an expected value of $\widehat{f}_n(X)$ with respect to the measure $f(x)dx$. For estimating this expected value, the arithmetic mean over all empirical means resulting from removing one of the data points of $\mathcal{D}$ leads to the so called 'leave-one-out' cross-validation estimator (LOO-CV) of risk $Q(m)$ (see Equation (3.53) in [13], and (2.5) in [15]), given by

$$Q_U(m) = \int \widehat{f}_n^2(x)dx - \frac{2}{n}\sum_{i=1}^n \widehat{f}_{(-i)}(X_i), \qquad (3)$$

where $\widehat{f}_{(-1)}(x)$ is the histogram resulting from the data $\mathcal{D}$ of which the i-th point has been removed. It differs from the empirical risk only by an unknown constant, and so its minimum is also m^*. The index 'U' stands for 'unbiased'.

In our case, the sample points X_i are not deterministic numbers (dates), but rather a set of uniformly distributed random variables. We therefore suggest to modify (3) by exchanging the arithmetic mean of $\widehat{f}_{(-i)}$ by the mixture of evaluations at uniformly distributed r.v.'s X_i across $[s_i, t_i]$:

$$Q(m) = \int \widehat{f}_n^2(x)dx - \frac{2}{n}\sum_{i=1}^n \frac{1}{\mu([s_i,t_i])}\int_{s_i}^{t_i} \widehat{f}_{(-i)}(x)dx, \qquad (4)$$

Also, the LOO-CV estimator of rist in our case becomes

$$\widehat{f}_{(-1)}(x) = \frac{1}{n-1}\sum_{j\neq i}^n \frac{\mu(B_k \cap [s_i,t_i])}{\mu([s_i,t_i])}. \qquad (5)$$

The change induced in the risk function by the added uncertainty of $\mathcal{D}$ consisting of random variables instead of fixed numbers is a reduction in the variance $\int \mathbb{E}[(f(x) - \widehat{f}_n(x))^2]dx$ of the histogram, and an increase in the bias $\int \mathbb{E}[f(x) - \widehat{f}_n(x)]dx$ ('more smoothing'). Since the risk is the sum of the (squared) bias error and variance, and since with a decreasing number of bins the bias error is increased and the variance is reduced, the effect of the switch to elements of $\mathcal{D}$ being random variables instead of numbers is a *reduction* in the number of bins (see [12], Sect. 4.1).

Evaluating (3) by inserting $\widehat{f}_n$ from (1) gives

$$Q(m) = \frac{2(n-1)}{h\cdot n^2}\sum_{i=1}^n \frac{\mu(B_k \cap [s_i,t_i])^2}{\mu([s_i,t_i])^2} - \frac{n+1}{h(n-1)}\sum_{j=1}^m w_j^2, \qquad (6)$$

which is compared and similar to $Q_U(m)$ - see Equation (6.16) in [12] and (3.53) in [13]. Note that the bin size is equal to $h = T/m$ for all bins.

Finally, for the determined histogram the clusters are defined as the intervals in time t, for which - uninterruptedly - $\widehat{f}(t) > p$. The parameter $p > 0$ indicates the minimal occurrence density for events in time to be qualified as being part of a cluster. In other words, the clusters are the connected components of

$$\widehat{f}^{-1}([p,\infty)) = \{\, t \in [t_0,T] \mid w_j > h\cdot p \text{ for some } j, \text{ where } t \in B_j\}. \qquad (7)$$

Writing this set as the disjoint union $\cup_k C_k$ of its connected components (intervals!) C_k, it is checked whether the resulting clusters have a credible amount of elements, i.e., whether for each k it holds $\#(C_k \cap \mathcal{D})/\mu(C_k) > p$.

Algorithm 1. Semi-parametric Histogram Clustering

Require: $h_0 > 0$; $\mathcal{H} \subset \mathbb{R}_+$; $t_0 \in \mathbb{R}$; $T \in \mathbb{R}_+$; $\mathcal{D} = \{X_i\}_{i=1}^n \subset \mathbb{R}$; $p > 0$
$\mathcal{H} = \{h > 0 \mid m \cdot h = T \text{ for some } m \in \mathbb{N}\}$
$h \in \mathcal{H}, m = T/h$
For $j \in \{1, \ldots, m\}$ let
   · $B_j \leftarrow [t_0 + (j-1)h, t_0 + j \cdot h)$
   · $\mathcal{D}_j \leftarrow B_j \cap \mathcal{D}$
   · $w_j := \sum_{i=1}^n \frac{\mu(B_j \cap [t_i, s_i])}{|t_i - s_i|}$ with $\mu([a,b]) = b - a$
$\widehat{f}(x) \leftarrow \sum_{i=1}^n \frac{w_i}{h} \mathbf{1}_{B_i}(x)$
$h \leftarrow h_0$
while $\widehat{J}(h)$ not minimal choose next $h \in \mathcal{H}$ and **do**
   $Q(m) = \frac{2(n-1)}{hn^2} \sum_{i=1}^n \frac{\mu(B_k \cap [s_i, t_i])}{\mu([s_i, t_i])} - \frac{n+1}{h(n-1)} \sum_{j=1}^m w_j^2$
   $m^* = \mathrm{argmin}(Q(m) \mid h = T/m \in \mathcal{H})$
   $\mathcal{C} = \widehat{f}^{-1}([p, \infty))$
   $\dot{\bigcup}_{k=1}^K C_k = \mathcal{C}$ with $C_k = [s_k^*, t_k^*]$ the connected components of $\mathcal{C}$.
   Check if for each $k \in \{1, \ldots, K\}$ it holds $\frac{\#(C_k \cap \mathcal{D})}{\mu(C_k)} > p$:
   if not **then**
      Remove h from $\mathcal{H}$.
      if $\mathcal{H} \neq \emptyset$ **then**
         rerun this loop with new $\mathcal{H}$.
      else
         There are no clusters.
      end if
   else
      $\mathcal{C}$ is the resulting clustering.
   end if
end while

Note that the minimizer of $m \mapsto Q(m)$ doesn't need to be the solution, as the clustering may still be inadequate if the clusters are underpopulated.

3 Application to Wooden Polychrome Sculpture

3.1 The Data

In the research project [3], the polychromy of medieval wooden sculptures have been investigated. The data reports includes 105 sculptures dated from the tenth up to the 13th century, with 480 categories. From these, the information on the use of translucent color on leaf metal and opaque paint layers have been extracted.

3.2 The Experiment

Figure 1 shows the application of the method to the data on translucent coloring of leaf metal sculpture's parts. The threshold p determining the minimal occupancy rate of a bin belonging to a cluster is indicated by the dotted horizontal line. Small bin sizes lead to strong inter-bin variances, and to small scattered clusters (Figs. 2 and 3).

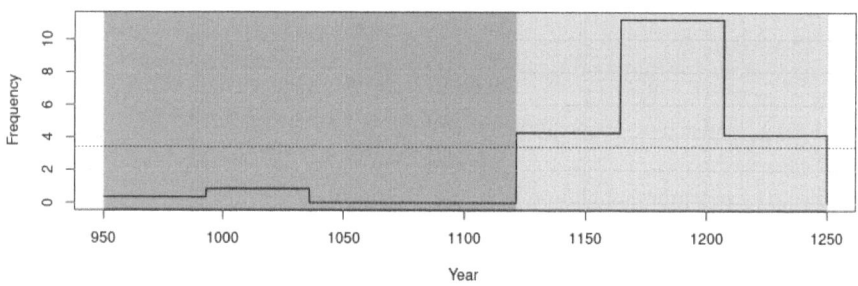

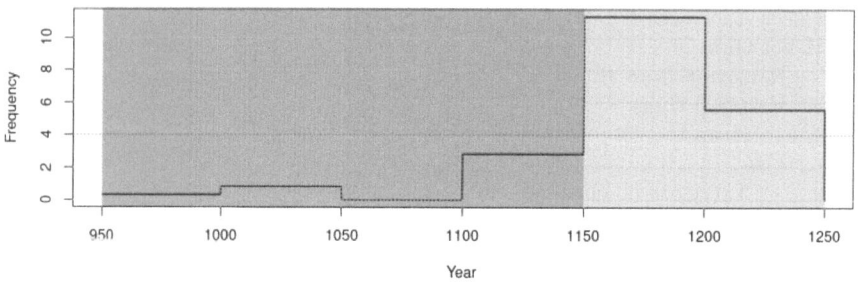

Fig. 1. Top: Translucent Color on Metal Leaf - there is a cluster of accumulated events of use of a translucent layer of color on metal leaf between 1120 and 1250. Bottom: Opaque Color - The period of accumulated use of translucent layers of color on opaque paint starts later, in 1150. Resulting histograms and thresholds for same occupancy threshold $p = 4.0$ per 50 years. Note that this results in the threshold of the (upper) histogram be below 4 (at 42.86), as the Frequency scale in both diagrams refers to number of cases per 50 years. (Color figure online)

From the initial choice of the number of bins and according bin-width we see that the choice with the smallest MISE ($m = 7$) removes the 'deep valleys' between the initally occuring clusters (top diagam in Fig. 1). The smoothing effect is strongly due to the choice of weights w_j being proportional to the ratio $\mu(B_j \cap [s_j, t_j])/\mu([s_j, t_j])$ which distributes the contribution of each item in $\mathcal{D}$ across the bins.

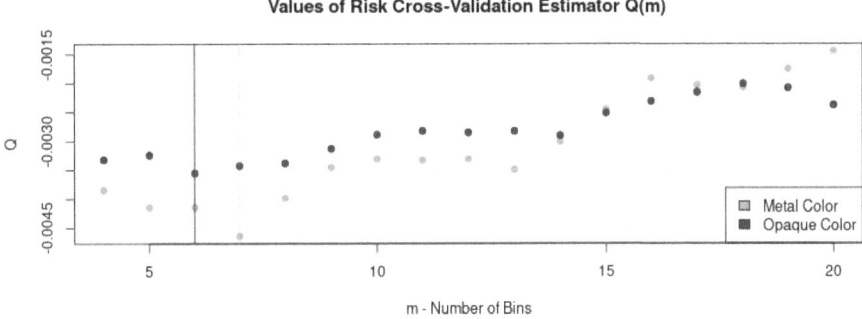

Fig. 2. The calculated values of the LOO-CV estimator of risk for metal and opaque colors: It is seen that the slightly richer metal data set leads to a slightly smaller bin size ($m^* = 7$) than the opaque color data ($m^* = 6$). The indicated threshold in either case corresponds to (a minimum of) 4.0 events per 50 years. (Color figure online)

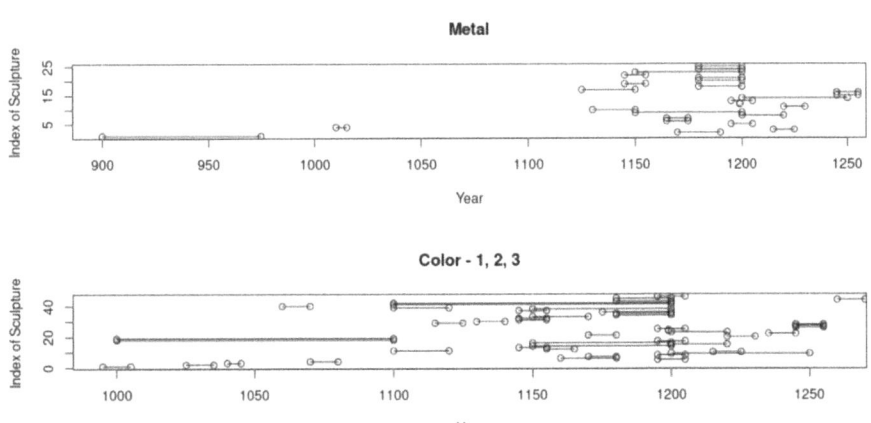

Fig. 3. Diagram of the available period data from the data sets 'Metal Color' (Top) and 'Opaque Color' (Bottom): The lines indicate the ranges of the sculptures' potential exact dates. Color 1-2-3 refers to the presence of three different (opaque) colors (green, red, blue), while the results on 'Opaque Color Red' in Table 1 refer to the corresponding subset of the shown data. (Color figure online)

Table 1. Results: Type of Polychromy, Threshold, Number of resulting Clusters, and Epochs. In the range of thresholds of [2.5, 4.0] the same number of clusters resulted for the metal and opaque color data sets, while the beginning of the resulting epoch varies moderately with the threshold. The restriction on data with red (opaque) color shows a second cluster for the smaller threshold.

Use of Translucence	p Min. Occup. per 50 years	#(Clusters)	Epochs
On Metal	4.0	1	1122-1250
On Metal	2.5	1	1000-1250
On Opaque Color	4.0	1	1150-1250
On Opaque Color	2.5	1	1100-1250
Opaque Color Red	4.0	1	1150-1250
Opaque Color Red	2.5	2	**1000-1050**, 1100-1250

We see that the same threshold for the resulting histogram results in also a single cluster, however, with a smaller extension, only starting in 1150, later than the beginning of the cluster of the metal coatings, however also reaching until the end of the observed interval (1250). If the threshold is lowered to a frequency of 2.5 counts per bin, the cluster for the Opaque Colors data increases to one with starting date 1100. For this threshold, the Metal data receives a cluster dating from 1000 to 1250.

As for restrictions (subsets) of the opaque color data to various specific colors (such as the result on 'red'), we find, that different cluster structures emerge, even clusters (on the time-axis) which are not subsets of the clusters from the original data set (see Table 1). Such a behavior is possible, when the data-subset changes the concentration of the strongly accumulated dates to a different epoch.

4 Discussion, Conclusion and Outlook

We mention that the method of Rudemo is appropriate in our case, because a *single sample* is given (see the discussion in [13], Chap. 3.2.2). The distribution being a mixture of uniform distributions of variable support required our modifikation of the weights w_k in the suggested way (equivalent to making a non-trivial choice of $\widehat{f}(x, X)$ in Equation (1.2) of [15]).

As to answer Questions 1. and 2. of Sect. 1.2, we state that the accumulation of the technique to use translucent colors can both be observed on the basis of an optimally and non-parametrically chosen histogram, with cluster-inducing thresholds based on expert-opinion. Various choices of these parameters yield the observation that invariably the application of a translucent cover to leaf metal predates the corresponding variant of this technique to opaque paint layers. Additionally, both techniques were employed concurrently with notable frequency between the years 1165 and 1200.

Concerning Question 3., applying the clustering only to the Opaque Color data for the color 'Red', we observe that for high thresholds (p = 4.0) the same cluster emerges, while for lower values, two clusters emerge, one which also starts already at 1000.

We conclude that by using the LOO-CV clustering method, a Knowledge-Guided semi-parametric clustering of data from art history allows the identification of periods of accumulated occurrences, even if onnly imprecise dating information (intervals) is available in the data set (as opposed to exact dates).

Restricting searches to sub-data sets corresponding to specific events allows identifying the presence of their accumulated occurrence (epochs). It is necessary, however, to perform the parameter-choice of the corresponding threshold (minimal bin occupancy) by guidance from domain-specific knowledge. In particular, the parameter p must be chosen under awareness of the richness of the considered data base in terms of the considered events. In our use case this is the knowledge about the data base on the polychromy of wooden sculptures [3], from which educated judgements of accumulated occurrences within a specified period of time can be made.

From the perspective of the data science of cultural heritage, the application of the non-parametric part of the rather conventional technique of histograms to a clustering task is interesting, because the principle of the bias-variance trade-off yields a natural way of choosing a resolution on the time-axis. Furthermore, the use of LOO-CV estimators in the context of recognizing the frequency of events suggests looking at their eminence in the sense what happens if they hadn't occurred (i.e., if they had been 'left-out'). This can be compared to modern machine learning notions, such as the 'counterfactuals' in causal inference [19], or local surrogate modeling in which the effect of a feature is measured by the change in parameters if its effect in the data set is corrected, i.e. artificially removed [20] (particularly (3) of [21], and (3) of [22]): The 'leave-one-out' estimators considered here also assess the effect of the removal of elements of the sample.

While this work demonstrates the technical applicability of LOO-CV histogram estimation for clustering of accumulated event data, our outlook is to compare the approach with standard scientific methods for the identification of art historical epochs [5].

Acknowledgments. E.S.'s work was supported by the Austrian Science Fund (FWF): P 32716-G "The Polychromy of Early and High Medieval Wood Sculpture", PI Elisabeth Sobieczky. F.S.'s work was supported by the FFG BRIDGE 'inAIco' project (FFG No. 878641) and the Austrian Ministry for Transport, Innovation and Technology, the Federal Ministry of Science, Research and Economy, and the Province of Upper Austria in the frame of the COMET center SCCH.

Disclosure of Interests. There are no competing interests to declare in association with this article.

References

1. Sobieczky, E.: Throne of gold and dress of stars. On the meaning of polychromy in high medieval marian sculpture. In: Znorovszky, A.-B., Jaritz, G. (eds.) Marian

Devotion in the Late Middle Ages (Studies in Medieval History and Culture), pp. 6–30. Routledge (2022). https://doi.org/10.4324/9781003179054-2
2. Sobieczky, E.: Pictura translucida between material and immaterial. Observations on high medieval european polychrome wood sculptures. In: Silva Santa Cruz, N., García García, F., Rodríguez Peinado, L., Romero Medina, R. (eds.) (In)materialidad en el arte medieval (Piedras Angulares), pp. 113–130. Trea, Gijón (2023). https://doi.org/10.21937/k6c1-6887
3. FWF (Austrian Science Fund) Principal Investigator Project P 32716-G "The Polychromy of Early and High Medieval Wood Sculpture", PI E. Sobieczky. www.akbild.ac.at/en/research/projects/research_projects/2019/the-polychromy-of-early-and-high-medieval-wood-sculpture
4. Albarakati, R.: Density Based Data Clustering. Electronic Theses, Projects, and Dissertations 134 (2015). https://scholarworks.lib.csusb.edu/etd/134
5. Moreno, H., Mendoza, A.S., Talavera, J.M., González, J.: Formation of clusters in cultural heritage-strategies for optimizing resources in museums. J. Cult. Heritage Manag. Sustain. Dev. (2021). ISSN 2044-1266
6. Ng, E.K.K., Fu, A.W.-C., Wong, R.C.-W.: Projective clustering by histograms. J. IEEE Trans. Knowl. Data Eng. **17**(3) (2005). https://doi.org/10.1109/TKDE.2005.47
7. Khachatryan, A., Müller, E., Stier, C., Böhm, K.: Improving accuracy and robustness of self-tuning histograms by subspace clustering. IEEE Trans. Knowl. Data Eng. **27**(9) (2015)
8. Kasper, S., et al.: Practical recommendations for the management of treatment-resistant depression with esketamine nasal spray therapy. World J. Biol. Psychiatry (2020). https://doi.org/10.1080/15622975.2020.1836399
9. Ulicsni, V., Babai, D., Vadász, C., Vadász-Besnyöi, V., Báldi, A., Molnár, Z.: Bridging conservation science and traditional knowledge of wild animals. MBIO J. Hum. Environ. (2018). https://doi.org/10.1007/s13280-018-1106-z
10. European Food Safety Authority. Outcome of the public consultation on the draft guidance on expert knowledge elicitation in food and feed safety risk assessment. EFSA **11**(6) (2014). https://doi.org/10.2903/sp.efsa.2014.en-544
11. Feigenwinter, L., Vetsch, D., Kammerer, S., Kriewitz, C., Boes, R.: Conceptual approach for positioning of fish guidance structures using CFD and expert knowledge. Sustainability **11**(6), 1646 (2019). https://doi.org/10.3390/su11061646
12. Wasserman, L.: All of Nonparametric Statistics. Pringer (2006). Chap. 6.1. ISBN-13: 978-0387-25145-5
13. Scott, D.W.: Multivariate density estimation: theory, practice, and visualization. Wiley Series in Probability and Statistics, 2nd edn. Wiley (1992) Chap. 3.2.1 & Chap. 3.3.2; ISBN: 0471547700,9780471547709
14. Sturges, H.A.: The choice of a class interval. J. Am. Stat. Assoc. **21**(153), 65–66 (1926)
15. Rudemo, M.: Empirical choice of histograms and kernel density estimators. Scand. J. Statist. **9**, 65–78 (1982)
16. Whittle, P.: On the smoothing of probability density functions. J. Roy. Stat. Soc. Ser. B (Methodol.) **20**(2), 334–343 (1958). https://www.jstor.org/stable/2983894
17. Birgé, L., Rozenholc, Y.: How many bins should be put in a regular histogram? ESAIM: PS **10**, 24–45 (2006). https://doi.org/10.1051/ps:2006001
18. Freedman, D., Diaconis, P.: On the histogram as a density estimator: L2 theory. Z. Wahrscheinlichkeitstheorie verw Gebiete **57**, 453–476 (1981). https://doi.org/10.1007/BF01025868

19. Pearl, J.: Causality. Cambridge University Press, Cambridge (2013). https://doi.org/10.1017/CBO9780511803161. ISBN 9780511803161
20. Fischer, L., et al.: AI system engineering-key challenges and lessons learned. Mach. Learn. Knowl. Extr. **3**(1), 56–83 (2021). https://doi.org/10.3390/make3010004
21. Neugebauer, S., Rippitsch, L., Sobieczky, F., Geiß, M.: Explainability of AI-predictions based on psychological profiling. Procedia Comput. Sci. 180, 1003-1012 (2021). https://doi.org/10.1016/j.procs.2021.01.361. ISSN 1877-0509
22. Sobieczky, F., Geiß, M.: Explainable AI by BAPC - Before and After correction Parameter Comparison (2023). https://arxiv.org/pdf/2103.07155

Author Index

A
Aslam, Uswa 29
Azeem, Muhammad 14

B
Bringas, Pablo G. 102

C
Cazzorla, Davide 91

F
Fischer, Lukas 67
Fuchs, Magdalena 67

G
Göhner, Ulrich 77
Gül-Ficici, Sebnem 77

I
Inayat, Irum 29

K
Khan, Saif Ur Rehman 14, 29
Kumar, Mohit 67

L
Laña, Ibai 102
Lunglmayr, Michael 57

M
Mashkoor, Atif 14, 29
Mencar, Corrado 91
Montuoro, Alessio 67
Moser, Bernhard A. 57, 67

N
Nisa, Habib Un 14, 29

P
Paudel, Sarita 44

R
Rey-Arnal, Danel 102
Riegler, Michael 3

S
Sametinger, Johannes 3
Schönegger, Christoph 3
Seliger, Raphael 77
Shaaban, Abdelkader Magdy 44
Sobieczky, Elisabeth 115
Sobieczky, Florian 115
Stadler, Marco 3

Y
Yousafzai, Abdullah 14

SPRINGER NATURE

GPSR Compliance

The European Union's (EU) General Product Safety Regulation (GPSR) is a set of rules that requires consumer products to be safe and our obligations to ensure this.

If you have any concerns about our products, you can contact us on ProductSafety@springernature.com

In case Publisher is established outside the EU, the EU authorized representative is:

Springer Nature Customer Service Center GmbH
Europaplatz 3
69115 Heidelberg, Germany

The manufacturer's authorised representative in the EU is Springer Nature Customer Service Centre GmbH, Europaplatz 3, 69115 Heidelberg, Germany. If you have any concerns regarding our products, please contact ProductSafety@springernature.com

Printed and bound by CPI Group (UK) Ltd, Croydon, CR0 4YY

26/03/2026

02078963-0007